Practical Organic
Chemistry

Practical Organic Chemistry

A student handbook of techniques

J. T. Sharp, I. Gosney
University of Edinburgh

and

A. G. Rowley
Consultant in analytical chemistry

London   New York
CHAPMAN AND HALL

First published in 1989 by Chapman and Hall Ltd
11 New Fetter Lane, London EC4P 4EE
Published in the USA by Chapman and Hall
29 West 35th Street, New York NY 10001

© 1989 J. T. Sharp, I. Gosney and A. G. Rowley

Typeset in 11/12 Sabon by
Best-set Typesetter Ltd, Hong Kong
Printed in Great Britain by
T. J. Press (Padstow) Ltd, Padstow, Cornwall

ISBN   0 412 28230 5

British Library Cataloguing in Publication Data

Sharp, J. T. (John T.). 1939–
    Practical organic chemistry.
    1. Organic chemistry. Laboratory techniques
    I. Title   II. Gosney, I. (Ian), 1942–
    III. Rowley, A. G. (Alan G.), 1948–
    547'.0028

    ISBN 0-412-28230-5

Library of Congress Cataloging in Publication Data

Sharp, J. T. (John Traquair), 1939–
    Practical organic chemistry: a student handbook of
techniques / J. T. Sharp, I. Gosney, and A. G. Rowley.
        p.   cm.
    Bibliography: p.
    Includes index.
    ISBN 0-412-28230-5 (pbk.)
    1. Chemistry, Organic – Technique.   I. Gosney, I.,
    1942–   . II. Rowley, A. G., 1948–   .   III. Title.
    QD258.S57   1989
    547 – dc19

Contents

Preface

One of the very best things about organic chemistry is actually doing experimental work at the bench. This applies not only at the professional level but also from the earliest stages of apprenticeship to the craft as a student. The fascination stems from the nature of the subject itself, with its vast array of different types of reaction and its almost infinite variety of different chemical compounds. Each reaction and each new compound pose their own particular problems to challenge the skill and ingenuity of the chemist, whether working in a first-year teaching laboratory or at the frontiers of research.

This book is intended to provide basic guidance in the essential experimental techniques used in a typical undergraduate course. It gives concise coverage of the range of practical skills required, from first-year level when students may have no previous experience, up to final-year level when students are usually involved in more complex and demanding experimental work in supervised research projects.

Our objective was to produce a handbook of techniques that could be used with a variety of practical courses throughout a student's whole period of study. Those who run practical courses generally have strong feelings about what particular experiments or exercises are appropriate for their own students, and it is rare that a book of experiments suitable for one department is acceptable to another. However, there is a common body of techniques applicable to all courses, and we hope that this book will provide a useful source of information on the range of techniques that are an essential part of current chemical practice. We have included not only the classical and timeless methods, for example the purification of compounds by crystallization and distillation, but also more modern techniques, such as those required for working with air- and water-sensitive reagents, without which most recent advances in organic chemistry would not have been possible. Also included are the modern methods of preparative chromatography, such as the 'flash' and 'medium-pressure' techniques, and the instrumental forms of chromatography, gas−liquid chromatography (GLC) and high-performance liquid chromatography (HPLC), which play such a vital role in the monitoring and analysis of reactions.

A book of this length cannot be comprehensive, and, as with all practical teaching, some techniques are better demonstrated than described. To this end the advice and guidance of experienced instructors are essential in the application of the techniques described to particular compounds or reactions under study.

Since the book is intended for use at different levels, the various chapters are structured so that the early parts of each section concentrate on learning how to handle the equipment and on the basic aspects of the technique. The later parts are concerned with more advanced aspects, such as the optimization of operating conditions or parameters. Basic theory, of chromatography for example, is dealt with only at the level needed for effective practical work.

While intended mainly for undergraduates, it is our hope that this book will also be of value to more advanced students as a guide to basic experimental methods onto which they can graft the refinements, modifications and extensions necessary for particular areas of research.

We thank our many colleagues (past and present), research students and undergraduates for their invaluable advice as to what is good, effective and safe laboratory practice at the present time. In particular, we thank Dr David Reid (of the University of Edinburgh NMR Service) for his advice on the section on the preparation of samples for nuclear magnetic resonance (NMR) spectroscopy. Finally, we express the hope that many of you, about to come to grips with the challenges of practical organic chemistry for the first time, will get as much pleasure and satisfaction from it as we have.

Edinburgh
October 1988
J. T. S., I. G. and A. G. R.

Foreword

There were two aspects of the teaching of organic chemistry which first attracted me to the subject. First, the application of the concepts of reaction mechanism to the rationalization of experimental observation and, secondly and most importantly, the enjoyment and fascination in performing synthetic reactions leading to pure products.

Although the methods now used in the characterization of these products are largely instrumental, the overwhelming requirement of an organic chemist is still for the undertaking of experiments to make compounds, and to isolate them in pure form. In order to enjoy this aspect of the subject the chemist must become skilled in the art of practical organic chemistry. Sadly, with the decrease in practical content of many undergraduate courses, the necessary skills are harder to acquire. This book gives an excellent grounding in the experimental techniques required for practical organic chemistry from first year level up to Honours level and beyond, and does so with due reference to essential modern safety practice. There are several excellent books which emphasize a range of interesting preparations and reactions. However, they lack any in-depth treatment of the basic practical requirements for the efficient performance of reactions and the isolation of pure materials, both crucial aspects of study for beginners to master. The authors have produced a text which should find wide appeal as it can be used in combination with books dealing with standard preparations. Although sophisticated computer programmes are being developed for mapping out the reaction paths to be followed in synthesis, it must be remembered that the vast number of products sold by the chemical and pharmaceutical industries are pure compounds and not computer print-outs. Producing these compounds demands a high degree of skill from the professional chemist.

Thus a text such as this volume, which guides young chemists through the correct practical procedures, has an important role to play in training people in experimental methods and deserves a place alongside them on the laboratory bench.

R. Ramage
Forbes Professor of Chemistry
University of Edinburgh
October 1988

Acknowledgements

We thank J. Bibby Science Products Ltd for permission to use some of their product drawings of 'Quickfit®' standard taper glassware, 'Bibby' plastic joint clips and 'Rotaflo®' stopcocks in our diagrams of apparatus assemblies. We also thank the American Chemical Society, Marcel Dekker Inc., Aldrich Chemical Company Ltd, and John Wiley and Sons Inc. for permission to use various copyright items of text, diagrams or tables as indicated in the text.

Safety and supervision in the laboratory

KEY SAFETY PRECAUTIONS

1. Work in the laboratory only during approved hours when supervision is available.
2. Wear safety spectacles (or a face shield) AT ALL TIMES. (Those who wear contact lenses, read Section 1.3.1.)
3. Do not eat, drink or smoke in the laboratory.
4. If you are in any doubt about experimental procedure or safe practice, then consult your instructor before proceeding.

More detailed safety precautions are given in Section 1.3; these must be read before starting experimental work.

SUPERVISION

The techniques described in this book represent accepted experimental practice. However, it must be emphasized that they are general descriptions, and their application to a particular chemical reaction or to particular chemical compounds may require modifications, either to make them effective for that particular case or for reasons of safety. For this reason it is essential that undergraduates and other inexperienced workers carry out practical work ONLY under the supervision of qualified personnel with due regard to safety considerations* and legislation.

* See *Guide to Safe Practices in Chemical Laboratories* published by the Royal Society of Chemistry, London.

1 Introduction

1.1 THE RANGE OF EXPERIMENTAL TECHNIQUES

Much of practical organic chemistry is concerned with what is generally called 'preparative' or 'synthetic' work, in which the objective is to carry out a chemical reaction, or a series of reactions, to produce a particular chemical compound in a pure state in as high a yield as possible. Such processes form the basis of the highly successful chemical industry, producing an enormous number of chemicals ranging from the simple compounds used in plastics and polymers to the highly complex compounds used in medicine.

A synthetic exercise in the laboratory usually involves a sequence of three operations: (a) carrying out the chemical reaction, i.e. the conversion of the reacting compounds into products; (b) separation of the required product or products from solvents, by-products or inorganic materials; and finally (c) the purification and identification of the product. In general, the techniques (a) used in carrying out reactions are fairly straightforward and involve bringing the reactants together in appropriate amounts and applying a stimulus such as heat or light to bring about the reaction (Chapter 2). However, in recent years this area has become more demanding as progress in synthesis has seen the increasing use of highly reactive, air- or water-sensitive reactants, which must be used at low temperatures or under inert atmospheres.

The techniques in group (b) are often referred to as the 'work-up' methods by which the required product is isolated from the reaction mixture. In some cases this may be very easy, but generally it is a more complex area in which the experimenter has to use his knowledge and judgement to call on various combinations of techniques such as extraction, chromatography or crystallization to bring about the required result. In early undergraduate exercises the work-up methods are usually clearly specified, but in more advanced project work this is a major area of decision making (Chapters 3 and 4). Much of this book is therefore devoted to the methods used for separation and purification and includes sufficient theoretical background to give the user an appreciation of their potential and limitations.

In addition to these techniques, which are directly related to preparative work, the chemist often needs to monitor a reaction as it takes place to check how far it has progressed or, at the end, to measure accurately the yield or ratio of products. This is usually achieved by using one of the instrumental chromatographic methods such as gas–liquid chromatography (GLC) (Section 4.1.2) or high-performance liquid chromatography (HPLC) (Section 4.1.3) with which measurements can be made using only microgram or milligram samples of the reaction mixture.

When a compound has been obtained from a chemical reaction, it must then be identified. Many physical properties have traditionally been used to identify known compounds (those which have been made before and whose properties are recorded in the chemical literature), e.g. melting point, boiling point and refractive index. These are still very important both as means of identification and as criteria of purity, but in addition there is available a battery of spectroscopic methods that provide enormously powerful tools for the identification of both known and new compounds. The most useful are infrared, ultraviolet, nuclear magnetic resonance and mass spectroscopies. This book does not cover the interpretation of spectra – on which many excellent texts are available – but does include information on the preparation of samples for spectroscopy (Chapter 5).

1.2 GOOD LABORATORY PRACTICE

The key to success in practical work is thinking before doing. Practical work should never degenerate into an exercise in blind faith (in the instructions) and blissful ignorance (of the chemistry) typical of the 'cookery book' approach. When tackling an exercise where full instructions are given, you should first read them all the way through – preferably before coming to the laboratory class – and (a) make sure you understand the chemistry and the objectives of the exercise, (b) make sure you understand the experimental techniques you will be using – if not, read them up before you start – and (c) think through each operation before you do it and try to visualize what you will be doing both at a molecular and a manipulative level, and so anticipate any possible difficulties or safety hazards. Thinking means that you will travel more hopefully, arrive much more often, and even learn some chemistry along the way.

You will also find it much easier to do good chemistry if you are well organized and work in clean and tidy conditions. It is often necessary to work on several experiments (or parts of experiments) at the same time – for example, some reactions need to boil under reflux for several hours, and in that time you could be working on the identification of an 'unknown' compound or working up another reaction mixture. This becomes easier as you get more experience, but it is important not to undertake so much that experiments are done in a hasty and slapdash way and not properly completed. This applies particularly in project work, where it is only too easy to get carried away on a wave of enthusiasm – remember that carrying out reactions is easy, it is the work-up and product characterization that takes most of the time.

However, best intentions and good instructions notwithstanding, there will be times when things will go wrong and occasionally times when nothing seems to go right. Usually it will be a 'hands-on' problem of not getting some experimental technique quite right because of lack of experience, or of not 'watching' a reaction carefully enough by monitoring its progress using thin-layer chromatography (TLC) or gas-liquid chromatography (GLC) (Sections 4.1.1 and 4.1.2). In project work it may even be that the theory is wrong.

1.3 SAFETY IN THE LABORATORY

SAFETY Key safety precautions are given before the Introduction (p. xiv).

Practical organic chemistry, when properly conducted, is a safe occupation, but care and forethought are needed to make it so. Many of the materials used in organic chemistry are flammable, or toxic in some way, or both. The development of sound working practices will ensure that such compounds can be used safely.

SAFETY All teaching institutions have their own safety precautions and procedures, and students should ensure that these are observed. Some general advice on common precautions is given below, but it is not comprehensive and should be supplemented by specific safety guidance for particular experiments.

1.3.1 Chemical hazards

In general, organic reactants selected for use in teaching laboratories should be of low toxicity but, even so, do follow these guidelines:

1. Keep ALL compounds and solvents away from the mouth, skin, eyes and clothes.
2. Avoid breathing vapours or dust.
3. Never taste anything in the laboratory.

Particular care is needed when working with strong acids, corrosive and volatile reagents and flammable solvents. In project work bear in mind that you may produce – by design or chance – new chemical compounds of unknown biological properties.

(a) Personal protection

(i) Eyes

SAFETY Safety spectacles (or a full face shield where greater protection is required) must be worn at all times while in the laboratory.

(ii) Contact lenses

Any student who wears contact lenses must take the most rigorous precautions to prevent any material entering the eye. Corrosive or toxic substances can rapidly penetrate behind the contact lens and irrigation (washing out) is almost impossible.

(iii) Hands

In general, careful manipulation and good practice should ensure that you keep chemicals off your hands. However, when using noxious, corrosive, or toxic materials it is sensible to use protective gloves, but bear in mind that these will make you clumsier at manipulation.

(iv) Clothes

A properly fastened laboratory coat is essential. It will provide some personal protection and avoid contamination of everyday clothing. Laboratory coats should be laundered regularly (with appropriate precautions if they are contaminated).

(b) General precautions

1. Do not heat, mix, pour or shake chemicals close to the face. Always point the mouth of a vessel away from the face and body.
2. Never pipette by mouth, always use a pipette filler.
3. Be careful with strong acids and alkalis, especially when heating. Never add water to concentrated acids or alkalis.

4. Materials giving off noxious fumes should be handled only in a fume-cupboard, wearing protective gloves. These include phosphorus halides, bromine, all acid chlorides, acetic anhydride, fuming nitric acid, concentrated ammonia solution, liquid ammonia, sulphur dioxide and others. If in doubt ASK your instructor.

(c) **Disposal of chemicals**

SAFETY Do not put organic solvents or any other organic materials down the sink.
Waste solvents should be placed in the receptacles provided and other residues disposed of as instructed.

1.3.2 Fire hazard

Most organic solvents and many other organic liquids are both volatile and flammable. Some form explosive peroxides when in contact with air (see item 5 below). General precautions to avoid fires are as follows:

1. Never heat organic liquids, even in small quantities, with or near a flame. Always use a water bath (Section 2.1.4), an oil bath (Section 2.1.4) or an electric heating mantle (Section 2.1.4). Particular care is needed with ether, light petroleum and carbon disulphide, which are very volatile and have low flash points.
2. Never heat organic liquids in an open vessel. A condenser must be used, either set up for reflux (Section 2.1.4) or distillation (Section 3.4.2). Some work-up instructions require the removal of a solvent from a reaction product by 'evaporation' – this requires the use of either a rotary evaporator (Section 3.1.2) or distillation (Section 3.4), NEVER direct evaporation into the atmosphere.
3. Never heat a closed system of any kind.
4. Before using ether (or any other volatile, flammable solvent) – for example, for extractions (Section 3.1.3) – make sure there are no flames or other sources of ignition (yours or your neighbour's) in the vicinity. It is often safer to work in a fume-cupboard than at the bench.
5. Some solvents, notably ethers and hydrocarbons, form explosive peroxides spontaneously on storage. Distillation of peroxidized solvents is highly dangerous, as the peroxide residues may explode violently when heated. Solvents of this type (check with your instructor) should therefore NEVER be evaporated or distilled un-

less a test has shown that peroxides are absent. A number of tests for peroxides and methods for their removal are given in references 1–3 on p. 91.

1.3.3 Vacuum and pressure work

1. Vacuum desiccators must be kept in a safety cage while under vacuum.
2. Do not evacuate flat-bottomed flasks, except Büchner flasks.
3. All vessels under pressure or vacuum should be kept and operated behind a safety screen. Do not use vessels that are scratched.

1.4 KEEPING RECORDS

The production of an adequate written record is a vitally important part of all experimental work. The final report should be accurate, clear and concise, and should contain enough information for any professional chemist to be able to replicate the work exactly. The established conventions and practices are set out in the guidelines below.

1.4.1 Recording experimental data

Keep all records in a robust laboratory notebook. Each exercise should be headed by an experiment number, the title and the date. During the course of the experiment enter all observations, weighings, melting points and other data directly into the notebook (do not write them on scraps of paper, which can be easily lost).

1.4.2 Final reports

When the experiment has been completed the final report should be written in the passive voice (as illustrated below) and should include:

1. A brief statement of the objective of the experiment.
2. A concise account in your own words of the experimental procedure actually used – do not simply copy out the directions given. Quantities of materials are placed in brackets after the name. An example is as follows:
   'Dry magnesium turnings (0.45 g, 0.018 mol) were placed in an oven-dried 25 ml three-necked flask equipped with a dropping funnel and reflux condenser, both fitted with calcium chloride tubes, and a magnetic stirrer. A solution of bromobenzene

(2.65 g, 0.017 mol) in dry ether (9 ml) in the dropping funnel was added over *c*. 5 min with stirring. After the first few drops had been added the solution became cloudy and began to warm up. The addition was then continued at a rate such that the ether boiled gently.'

   Detailed descriptions of standard experimental procedures such as distillation or crystallization are not generally required (except for experiments specifically designed to teach such techniques), but do include a note of any variations that were important for the particular experiment.

3. The weight of each product and its percentage yield:

$$\text{yield (\%)} = \frac{\text{yield obtained}}{\text{theoretical yield}} \times 100$$

4. The melting point or boiling point of each product together with literature values for comparison (obtainable from reference books in the laboratory or library: see Chapter 6).

5. Spectroscopic data on the products if required to determine their identity or purity. Infra-red (IR) and/or nuclear magnetic resonance (NMR) spectra are usually used to characterize known compounds, and again literature values should be given. Sources of literature data are given in Chapter 6. For IR spectra it is usual to quote only significant characteristic group absorptions, but for NMR spectra the full spectrum should be reported (both chemical shifts and coupling constants).

6. A concluding paragraph summarizing the results and commenting on them.

1.5 SAMPLES AND SPECTRA

Keep small samples of all products, intermediates and derivatives, and label the sample tube with your name, experiment number, date, compound name and its melting point (m.p.) or boiling point (b.p.). Spectra should be similarly labelled and in addition should have noted on them the conditions and instrument parameters under which they were run.

2 Carrying out reactions

For a preparative process (A + B → C) the basic requirement is to carry out the reaction under conditions in which the starting materials A and B will react at a convenient rate, with the minimum of side-reactions, and under which the product C is stable. The conditions required vary enormously with the nature of the reaction and primarily involve control of the reaction temperature, the way in which the reactants are mixed together and, in some cases, protection of the reaction mixture from atmospheric oxygen and water.

In preparative work carried out in a teaching laboratory, particularly in the early stages of training, the reaction conditions will usually be specified in some detail in the instructions for the experiment. It will not therefore be necessary to make decisions about the reaction temperature, which solvent to use, and how long a reaction time is required. However, in more advanced work and particularly in project work this becomes a critical area of decision making which requires careful thought about the nature of the chemical reaction in hand and the exercise of the experience and expertise built up in earlier work.

In project work the importance of monitoring the progress of reactions cannot be overemphasized. This usually involves the use of one of the analytical methods of chromatography (Section 4.1) to follow the disappearance of the reactants and/or the formation of the product as the reaction is going on. A little time spent in setting up a monitoring method is usually amply repaid in time saved by not having to repeat reactions and by knowing what is present in the reaction mixture when it comes to devising the best 'work-up' method for obtaining the product.

2.1 BASIC TECHNIQUES

2.1.1 Apparatus

Laboratory-scale chemical reactions are usually carried out in boro-silicate glass (e.g. Pyrex®) apparatus interconnected by tapered ground-glass cone-and-socket joints (Fig. 2.1). The reactions are usually carried out in round-bottomed (RB) flasks (Fig. 2.2) with single or multiple necks to permit the attachment of condensers, dropping funnels, stirrers, etc., as shown for example in Fig. 2.3. Single-neck flasks can be modified using two- or three-necked adapters (Fig. 2.4).

The cone-and-socket joints come in various sizes and a range of reduction and expansion adapters (Fig. 2.1) is available for inter-connecting pieces of apparatus with different joint sizes. For most purposes the joints can be assembled 'dry' (without joint grease), when a slight push and twist on assembly will provide a weak fric-tional locking effect. The various components of an assembly of ap-paratus must be supported by retort-stand clamps as common sense dictates to prevent the joints pulling apart. Plastic spring clips (Fig. 2.1) can be used to hold the two parts of the joint together when necessary, e.g. at the flask/vapour duct joint in the rotary evaporator (Fig. 3.1).

When used dry, the joints provide a good seal for liquids but the liquid does permeate into the ground-glass interface and some solu-

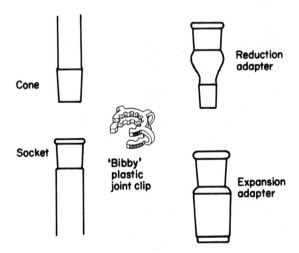

Fig. 2.1 Cone-and-socket (standard taper) ground-glass joints for intercon-necting glass apparatus.

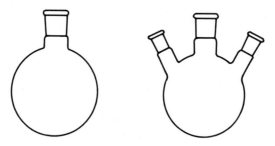

Fig. 2.2 Round-bottomed (RB) reaction flasks.

tions – notably aqueous sodium hydroxide – can cause irreversible seizure of the two parts. This can be prevented by using a thin Teflon® sleeve over the cone or a sparing application of joint grease (either hydrocarbon (Apiezon) or silicone grease). Joints must also be lightly greased when it is necessary to make them gas-tight to prevent the

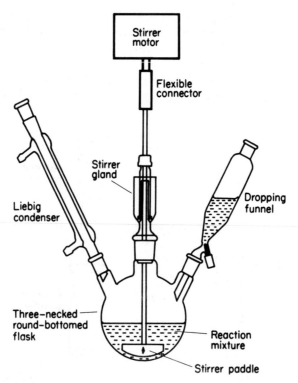

Fig. 2.3 A typical assembly of apparatus for preparative work.

Fig. 2.4 Two- and three-neck adapters.

ingress of air (Section 2.1.5) or when rotational movement of the two parts is required.

2.1.2 Addition of reactants

Reactions can be carried out in several different ways depending on their nature: (i) in some cases appropriate amounts of the reactants are weighed out and the whole amounts are mixed together in the reaction vessel before the start of the reaction; (ii) more often the whole amount of one of the reactants is placed in the reaction vessel and the other is added gradually over a period as the reaction progresses; and (iii) in rarer cases both reactants are added gradually during the reaction.

(a) Weighing and transfer*

In some cases – in small-scale reactions – the reactants can be weighed out directly into the reaction vessel, but in general it is preferable to weigh them out into separate containers and then transfer them. Solids are most easily weighed out in a beaker covered with a watch glass and transferred to the reaction vessel using a powder funnel (Fig. 2.5) to avoid contaminating the ground-glass socket. If a solvent is to be

Fig. 2.5 A powder funnel.

* For air- or water-sensitive materials, see Section 2.2.1.

used it is often convenient to dissolve the solid before transfer. Liquids may be weighed out in stoppered conical flasks or, more conveniently, dispensed by volume (if the density is known) using a measuring cylinder. Again a funnel should be used for transfer to the reaction vessel or the dropping funnel.

(b) Slow addition of reactants

(i) Liquids and solutions

In moderate- and large-scale reactions these can be dripped in from a dropping funnel (Figs. 2.3 and 2.6)*. Glass stopcocks should be lightly greased for easy control of the drip rate but be careful not to use excess grease, which may plug or partially plug the hole in the stopcock (and later dissolve out with consequent change to the drip rate). 'Rotaflo®' stopcocks have only Teflon® parts in contact with the liquid and require no grease.

(ii) Solids

The gradual addition of solids is not easy and they are best added as solutions when this is acceptable. If not, then the solid may be added batchwise via a powder funnel. However, in the majority of cases

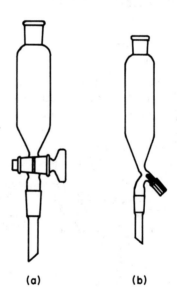

(a)   (b)

Fig. 2.6 Dropping funnels with (a) glass and (b) 'Rotaflo®' stopcocks.

---

* Syringe techniques for the addition of small volumes are discussed in Section 2.2.1.

where the reaction is being done in a boiling solvent, it will be necessary to allow the reaction vessel to cool briefly before each addition.

2.1.3 Stirring reaction mixtures

Stirring a reaction mixture is often necessary to provide good mixing as reactants are added, to keep solids or oils in suspension, or to promote smooth boiling for reactions under reflux (Section 2.1.4). There are two main methods: (a) a paddle stirrer on a shaft directly connected to a stirred motor (Fig. 2.3), and (b) a magnetic stirrer bar driven by a rotating magnet mounted below the reaction vessel (Fig. 2.7).

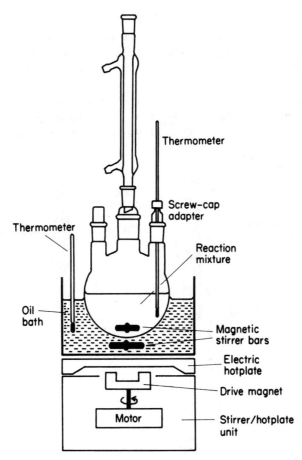

Fig. 2.7 The use of a stirrer/hotplate and oil bath for heating and stirring a reaction mixture.

Shakers – in which the whole vessel and its contents are shaken – are rarely used in preparative work. However, they are occasionally useful – for example when the violent agitation of a two-phase mixture is required as in phase-transfer reactions.

Paddle stirrers are more trouble to set up than magnetic stirrers (see below) but are essential for large-scale reactions and, on any scale, for viscous solutions or those containing much suspended solid material, when magnetic stirrers are not effective. In choosing the type to use, remember that in some reactions solids or complexes are formed and the mixture becomes more difficult to stir as the reaction proceeds. If in doubt, use a paddle stirrer.

(a) Paddle stirrers

The most effective stirrers of this type utilize a Teflon® (polytetra-fluoroethylene or PTFE) paddle shaped to fit the flask and mounted (detachably) on a glass driveshaft (Fig. 2.8a). (Teflon® is inert to

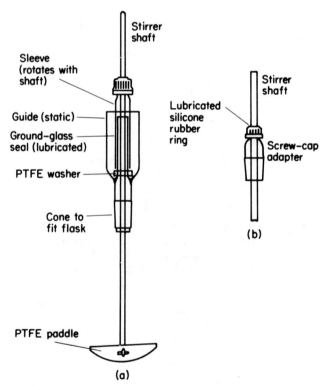

Fig. 2.8 (a) Paddle stirrer/sleeve gland assembly; (b) screw-cap adapter used as a simple stirrer seal.

everything except fluoride and molten alkali metals, and can be used at temperatures up to *c.* 250°C.) The stirrer guide (Fig. 2.8a), which fits into the neck of the flask, incorporates a ground-glass sleeve gland that is lubricated with viscous oil (a silicone oil or medicinal liquid paraffin) and provides an effective seal to prevent the loss of vapour from boiling solvents or the ingress of air or water vapour. The sleeve (the rotating part of the gland) is attached to the stirrer shaft by a screw-cap/silicone rubber ring compression fitting. For slow, short-duration stirring a screw-cap adapter (Fig. 2.8b) can be used as a stirrer guide/seal if the silicone rubber ring inside the cap is lubricated with liquid paraffin or silicone oil.

Some flexibility in the stirrer motor/driveshaft connection is desirable for ease of setting up and smoothness of running. This is usually achieved by using a short piece of thick-walled rubber tubing as the connector, held by two Jubilee clips or twisted copper wire (Fig. 2.3).

SAFETY Stirrer motors create sparks and should not be used where there are any flammable vapours. They normally have a built-in speed control but generally tend to increase in speed as they warm up. Adequate equilibration time should therefore be allowed before such a set-up is left unattended. Careful attention should also be paid to secure clamping of the apparatus as the vibration from the stirrer can cause clamps and cone-and-socket joints to work loose.

(b) Magnetic stirrers

As noted above, these are very convenient and effective for systems that are easy to stir. Stirrer bars are now almost invariably Teflon® coated (see (a) above for Teflon® limitations), but beware of the older, cheaper variety coated with other plastics that may not withstand some solvents or high reaction temperatures. Stirrer bars encapsulated in glass should be used for reactions involving molten alkali metals, which blacken Teflon®.

The magnet drive units are available as simple stirrers for reactions at room temperature or as very useful combined stirrer/hotplates (e.g. Fig. 2.7) or stirrer/heating mantles for reactions that require stirring and heating (see also next section).

(c) Shakers

The most common type has a motor-driven arm that moves up and down in an osillatory motion with a clamp at the end to hold the reaction vessel.

SAFETY Four points are important: (i) the stopper of the reaction vessel should be clipped or wired on; (ii) the shaker should be balanced by using a similar weight on the opposite arm; (iii) shakers tend to increase in speed as the motor warms up, so they should not be left until the required steady speed has been attained; and (iv) it is sensible to use the shaker behind a safety screen and in a fume-cupboard in case the flask should break.

2.1.4 Temperature control

(a) Heating reaction mixtures

Almost all preparative reactions are carried out in the liquid phase in a reaction solvent of some kind even in cases where the reactants themselves are liquids. The boiling point of the solvent is particularly important since this provides the most convenient way of controlling the reaction temperature (see (i) below). In teaching exercises the solvent to be used will be specified in the instructions, but for project work some useful information on the properties of solvents and methods for their purification is given in the references listed on page 24.

(i) Boiling under reflux

Reactions carried out in boiling solvents (e.g. Figs. 2.3 and 2.7) utilize a condenser set up vertically to return the condensed solvent to the reaction vessel – a technique usually called 'boiling under reflux'. Anti-bumping granules or stirring should be used to promote smooth boiling.

 HEATING METHODS. Electric heating mantles (Fig. 2.9) provide the safest and most effective way of heating RB flasks under reflux conditions. For both effectiveness and safety reasons it is important to use a mantle of the proper size (flask capacity is always indicated on a plate on the mantle). The power input to the mantle should be adjusted to produce the required slow solvent reflux.

 Electrically heated water baths can be used for heating reactions under reflux when using a solvent of boiling point up to *c.* 80°C. In rare cases this may be preferable to using a heating mantle – for example where the reaction involves highly sensitive reactants or products and the surface temperature of the flask must be kept as low as possible. The limitations of water baths are obvious: (i) the risk of boiling dry makes them unsuitable for unattended or overnight reactions; (ii)

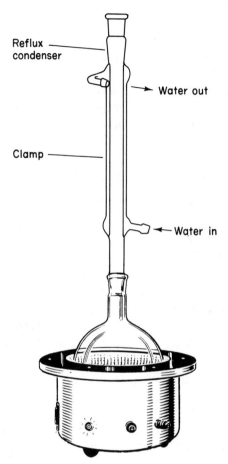

Reflux condenser

Water out

Clamp

Water in

Fig. 2.9 Electric heating mantle.

steam and condensation envelopes the apparatus, making anhydrous reactions more difficult to do; and (iii) they are not suitable for reactions involving sodium or other species that react violently with water because of the hazard if the flask should be cracked or a joint sprung during the operation.

CONDENSERS. A simple Liebig water-cooled condenser (Fig. 2.10) is adequate for liquids of boiling point above 50°C, but for those of lower boiling point, e.g. diethyl ether (b.p. 35°C), the more efficient double-surface condenser (Fig. 2.10) is required.

When fitting rubber tubing to a condenser, hold the condenser

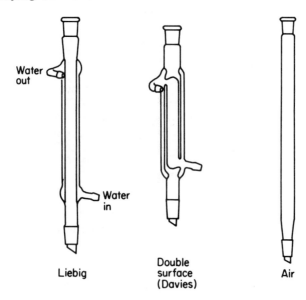

Water out

Water in

Liebig

Double surface (Davies)

Air

Fig. 2.10 Condensers.

with a cloth to protect your hand in case of breakage, and lubricate the tubing with water, ethanol (better), or a little joint grease. This problem is absent on the latest condensers with plastic screw-fitting water connectors.

SAFETY Condenser tubing should be secured to the condenser and water tap with copper wire (or tube clips) if the reaction is to be left overnight (but beware of perished rubber tubing, which can split when being wired on). It is also necessary to incorporate a water-operated cut-out switch into the power supply to the flask heater so that the power is cut off if the water supply fails.

Such potential problems can be avoided in some cases by using an 'Airflux'* condenser — a water-cooled condenser that needs no water supply. This device is a modified Liebig condenser whose cooling water is itself cooled by convected circulation through an aluminium heat sink. The heat capacity is limited, however, and these condensers are suitable only for liquids of boiling point in the range 60–150°C.

Air condensers (Fig. 2.10) may be used for high-boiling liquids (b.p. $\gtrsim$ 150°C).

* Jencons Scientific Ltd, Leighton Buzzard, Bedfordshire.

(ii) Reactions not under reflux

In some cases a reaction may require heating but there may not be an appropriate solvent for carrying it out under reflux at the temperature required. In such cases the reaction vessel is usually heated in an oil bath or (for high temperatures) a molten-metal bath.

OIL BATHS. A set-up (e.g. Fig. 2.7) in which the flask is heated in an oil bath by an electric hotplate is used for reaction temperatures from ambient up to c. 250°C. It is usual to use a combined stirrer/hotplate with a magnetic stirrer bar in the oil bath in addition to the one in the reaction flask. Alternatively, a paddle stirrer can be used to stir the reaction mixture if necessary (Section 2.1.3). The temperature of the reaction mixture is monitored using a thermometer held in a screw-cap adapter (Fig. 2.7). It is always essential to have another thermometer for monitoring the bath temperature.

The ease with which the temperature can be controlled depends much upon the degree of sophistication of the hotplate. The best types are those with thermostatic control and a temperature-sensing probe immersed in the oil. These provide accurate control at any selected temperature independent of the ambient conditions in the laboratory. Less expensive hotplates have a simple power controller, and when using these you must start at a low setting and work up gradually until the required oil temperature is reached. It takes some time for the system to come to equilibrium at each setting and it is easy to overshoot the required temperature. Temperature control with this type is less precise but quite adequate for most preparative reactions; however, changes in ambient temperature, changes in stirrer speed, or draughts in fume-cupboards can cause appreciable variations.

The 'oil' used in oil baths is commonly medicinal liquid paraffin, which is cheap and adequate for temperatures up to c. 200°C. Above this temperature it smokes, darkens rapidly and can catch fire. At temperatures above c. 150°C it is better to use the oil bath in a fume-cupboard because of the unpleasant 'hot-oil' smell and fumes. Silicone oils give a wider temperature range; for example Dow Corning 550 silicone fluid can be used from ambient temperature up to 250°C. However, these fluids are very expensive and are usually reserved for high-temperature use only.

The vessel used for the bath is usually a crystal dish (Pyrex®) of appropriate size for flasks up to 250 ml capacity. These dishes are shallow enough not to impede the clamping of the apparatus and have a flat base for good heat transfer from the hotplate. Larger flasks (500 ml and above) are best accommodated in aluminium saucepans (not enamelled iron, which blocks the action of magnetic stirrers).

SAFETY Oil baths that become contaminated with water can be dangerous when heated above 100°C and should be discarded or the water removed.

METAL BATHS. For higher temperatures a bath of molten Woods metal should be used. This is an alloy of lead, bismuth, tin and cadmium that melts at 70°C and can be used up to *c.* 350°C. It is generally used in an enamelled iron mug or saucepan and heated with a Bunsen burner. Since it expands on solidification, any glassware (e.g. thermometers) left immersed will be destroyed.

(b) Reactions at sub-ambient temperatures

An apparatus set-up similar to that in Fig. 2.7 is used but with the flask surrounded by a bath of coolant at the required temperature (e.g. Fig. 2.16, p. 30). A crystal dish may be used as the bath vessel down to *c.* −20°C, but below that a shallow, wide Dewar vessel (Fig. 2.16) is required. For good control, particularly at low temperatures, it is important to monitor the temperature of both the reaction mixture and the coolant bath. Normal mercury thermometers are usually calibrated down to −10 to −20°C, but for lower temperatures an alcohol thermometer (range −120 to +30°C) is required. Common cooling agents are discussed below.

(i) Ice

A slurry of crushed ice and water provides a constant 0°C. Lower temperatures can be achieved by using well stirred mixtures of crushed ice and inorganic salts, e.g. a 3:1 ratio of ice/sodium chloride will give temperatures down to *c.* −20°C. For other examples, see Table 2.1.

(ii) Solid carbon dioxide (Cardice)

This material can be used to cool baths down to −78°C. It is usually supplied in large blocks. Crushing is easily done by breaking off chunks from the block with a hammer or ice-pick, wrapping the chunks in a strong cloth and crushing them with a mallet or a block of wood. Frostbite is avoided by handling the Cardice rapidly using heavy rubber gloves.

Table 2.1 Cooling baths using ice/salt mixtures[a]

Salt	Concentration (g salt/100 g ice)	Temperature (°C)
KCl	30	−10
NH$_4$Cl	25	−15
NaCl	33	−21
NaBr	66	−28
MgCl$_2$	85	−34
CaCl$_2$·6H$_2$O	123	−40
CaCl$_2$·6H$_2$O	143	−55

[a] Drawn from a more extensive set of data in Gordon, A. J. and Ford, R. A. (1972) *The Chemists Companion*, Wiley-Interscience, New York. Reprinted with permission of John Wiley and Sons Inc.

SAFETY Baths containing Cardice should be prepared and used in a fume-cupboard because much CO$_2$ and solvent vapour can be given off, particularly in the preparation.

CARDICE/ACETONE (−78°C). Cardice is most often used in combination with acetone as the bath fluid to produce a constant-temperature bath of −78°C. The technique is to put the acetone into the Dewar vessel first and then add the crushed Cardice slowly (beware vigorous foaming in the early stages) until an excess is present.

CARDICE/OTHER SOLVENTS. Baths of a reasonably constant temperature above −78°C can be produced by adding a small excess of solid lumps of Cardice to organic solvents (or solvent mixtures) that have freezing points above −78°C[1]. This is a variation on the slush-bath technique described below. The following solvents are useful: carbon tetrachloride (−23°C), heptan-3-one (−38°C), cyclohexanone (−46°C) and chloroform (−61°C). Intermediate temperatures can be set up using mixtures of *o*- and *m*-xylene (Fig. 2.11).

(iii) Liquid nitrogen
Liquid nitrogen itself will provide cooling down to −196°C and is often used in combination with organic solvents to produce 'slush baths' at higher temperatures.

1. Phipps, A. M. and Hume, D. N. (1968) *J. Chem. Educ.*, **45**, 664.

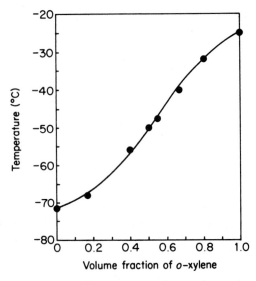

Fig. 2.11 Bath temperatures for mixtures of *o*- and *m*-xylene with Cardice.

SLUSH BATHS. The technique is to place the appropriate solvent (see selected examples in Table 2.2)[2] in a Dewar vessel and add liquid nitrogen very slowly with vigorous stirring (in fume-cupboard) until an appreciable part of the solvent has solidified. The resulting 'slush' will remain at the freezing point of the solvent as long as solid is present.

SAFETY Never mix liquid air or liquid oxygen with organic solvents as violent explosions may result.

2.1.5 Reactions under anhydrous conditions and inert atmospheres

Many organic reactants, intermediates and reaction solvents react readily with water and/or atmospheric oxygen or carbon dioxide. Thus preparative procedures must often be carried out under absolutely anhydrous conditions and with the exclusion of air. This section deals only with the basic techniques used for the type of reaction in which the reactants, solvent and apparatus must be dry for the reaction to take place (e.g. Grignard reactions, malonic ester syntheses, etc.) and where the reaction itself must be carried out with the

2. Rondeau, R. E. (1966) *J. Chem. Eng. Data*, **11**, 124.

Table 2.2 Slush baths with liquid nitrogen

Solvent	Temperature (°C)
cyclohexane	6
cycloheptane	−12
o-dichlorobenzene	−18
m-dichlorobenzene	−25
o-xylene	−29
bromobenzene	−30
chlorobenzene	−45
n-octane	−56
ethyl acetate	−84
heptane	−91
cyclopentane	−93
hexane	−94
toluene	−95
methyl acetate	−98
cyclohexene	−104
ethyl alcohol (viscous)	−116
n-pentane	−131
2-methylbutane	−160

Reprinted with permission from ref. 2. Copyright (1966) American Chemical Society.

exclusion of air and water vapour. (More advanced techniques for handling highly air- and water-sensitive reagents such as organo-lithium solutions are discussed in Section 2.2.1.)

(a) Reactions under anhydrous conditions

(i) Methods for drying reactants and solvents

In teaching laboratory preparations, the reactants and solvents will usually be dried before issue to save time. If not, then specific drying instructions should be given. A general outline of the methods used is given below for liquids and solids.

LIQUIDS. Organic liquids (reactants and solvents) are dried by quite specific methods depending on the nature of the compound. If details are not given, you should consult your instructor or project supervisor before proceeding. References 3–6 provide much information on the drying and purification of a wide range of solvents and reagents.

The method generally involves adding to the liquid a small amount of an inorganic drying agent that will either absorb water (e.g. molecular sieve, or anhydrous salts such as magnesium sulphate or calcium sulphate) or react irreversibly with water (e.g. sodium metal, calcium hydride or lithium aluminium hydride). Obviously the drying agent must not react with the compound. In some cases the mixture of drying agent and organic liquid are simply stirred at room temperature and then filtered to remove the drying agent – using a dry sintered funnel and receiver (p. 74). The liquid is then distilled by standard methods using dry apparatus. In other cases the organic liquid is boiled under reflux with the drying agent and then distilled from it. Standard reflux and distillation techniques can be used, but the apparatus shown in Fig. 2.12 allows both reflux and collection without disassembly of the apparatus. The tap at X is kept open during the reflux period and then closed to allow collection of dry distilled solvent in the bulb. It is particularly useful when a regular supply of a freshly distilled solvent such as tetrahydrofuran or dimethoxyethane is required.

SAFETY In the use of this apparatus and in any solvent distillation it is absolutely vital that a good supply of solvent is kept in the flask and that it is not distilled to dryness or near dryness. The apparatus should NOT be left unattended during distillation. For general comments on safety in solvent distillation, see Section 1.3.2 and 3.4.1.

SOLIDS. Drying methods for solids are less specific and involve the use of a desiccator (Fig. 2.13) for drying at room temperature or a drying pistol (Fig. 3.14, p. 77) or vacuum oven for drying at elevated temperatures. Organic solids should NEVER be dried by placing them in a normal laboratory oven.

The desiccant in a desiccator is kept in a separate dish below the

3. Fieser, L. F. and Fieser, M. (1967–88) *Reagents for Organic Synthesis*, vols. 1–13, Wiley-Interscience, New York.
4. Riddick, J. A. and Burger, W. B. (1970) *Organic Solvents: Physical Properties and Methods of Purification*, in *Techniques of Chemistry*, 3rd edn, vol. 2, Wiley-Interscience, New York.
5. Gordon, A. J. and Ford, R. A. (1972) *The Chemist's Companion*, Wiley-Interscience, New York.
6. Perrin, D. D., Armarago, W. L. F. and Perrin, D. R. (1980) *Purification of Laboratory Chemicals*, 2nd edn, Pergamon Press, New York.

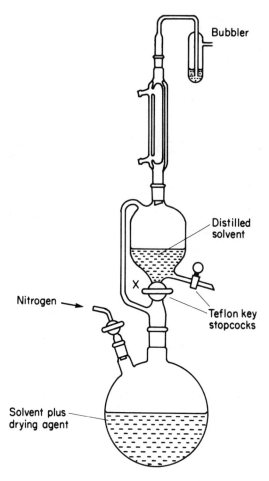

Fig. 2.12 Apparatus for drying and distilling small quantities of solvent.

metal gauze. Self-indicating silica gel is commonly used, which is blue when dry and pink when in need of regeneration (heat in oven at 125°C). For critical applications a more powerful desiccant such as phosphorus pentoxide should be used.

SAFETY Caution: phosphorus pentoxide reacts violently with water.

The solid to be dried should be in powder or fine crystalline form and should be spread out thinly in a crystal dish (covered loosely with a watch or clock glass when drying under vacuum). Drying is

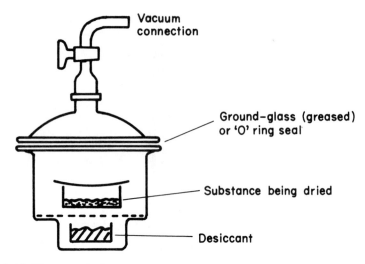

Fig. 2.13 Vacuum desiccator.

faster and more effective under vacuum and the desiccator can be evacuated to *c.* 10–12 mmHg using a water (filter) pump, see Section 3.4.8(a) or to *c.* 0.1 mmHg using a rotary oil pump, see Section 3.4.8(b).

SAFETY The desiccator must be placed inside a metal mesh guard cage before evacuation is started, in case of implosion, and kept there while under vacuum.

When letting air back into the desiccator, place a small scrap of filter paper over the end of the tube and then cautiously open the tap. The filter paper moderates the inrush of air, preventing the crystals and desiccant being blown about.

In some cases – for non-volatile solids – heating under vacuum may be required. This technique is more often used for drying reaction products than reactants and is discussed in Chapter 3.

(ii) Drying apparatus

Glassware can most easily be dried by baking in a laboratory oven. A period of 1–2 h at about 120°C is enough to remove surface water and will provide adequate drying for most teaching-laboratory purposes. However, for highest yields and where absolute dryness is required, the apparatus should be baked at 125°C overnight or at 140°C for 4 h to remove adsorbed water. Items containing close-fitting parts such as syringes or stirrer seals should be disassembled before baking. Don't forget to bake ancillary apparatus such as measuring cylinders, dropping pipettes, flasks, beakers and crystal dishes for measuring or weighing out the reactants. Plastics – including Teflon® – should not be baked but should be dried in a desiccator.

After baking, the apparatus should be assembled while hot (use gloves), using a little joint grease on all ground-glass joints to prevent seizure and to provide a good seal. The assembly should then be allowed to cool using drying tubes (see next section) on condensers, dropping funnels, etc., to prevent ingress of moisture (Fig. 2.14), or preferably while flushing with dry nitrogen (see subsection (b) below). Small pieces of apparatus, e.g. measuring cylinders, syringes, etc., should be allowed to cool in a desiccator.

(iii) Apparatus assemblies for anhydrous reactions

Normal preparative apparatus (e.g. fig. 2.14) can be easily modified for anhydrous working by: (a) the use of drying tubes (see below) on the condenser and dropping funnel, (b) the use of joint grease on ground-glass joints, and (c) the use of a magnetic stirrer or a well sealed gland on a paddle stirrer (Section 2.1.3).

DRYING TUBES. Drying tubes (Fig. 2.14) usually contain granulated self-indicating silica gel (blue when dry, pink when in need of regeneration) or granulated calcium chloride held between two plugs of glass wool.

SAFETY Use rubber gloves or tongs when handling glass wool.

The tubes should be baked at 125°C after charging with desiccant and cooled in a desiccator before use.

When absolute dryness is required it is best to establish both anhydrous and inert-gas conditions by flushing the system with a dry inert gas (usually nitrogen) as discussed in the next subsection.

(b) Reactions under inert atmospheres

(i) Supply of inert gas

Nitrogen is the most commonly used inert gas since it is cheap and readily available in a high standard of purity. ('White Spot' oxygen-free nitrogen from the British Oxygen Company contains only 2 ppm

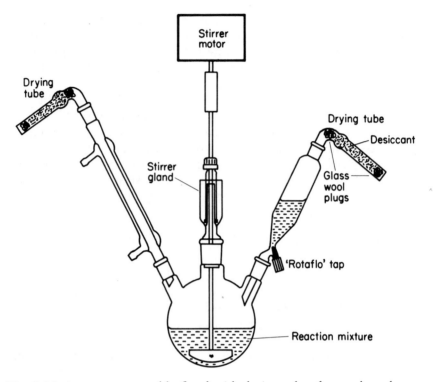

Fig. 2.14 Apparatus assembly fitted with drying tubes for work under anhydrous conditions.

(parts per million) of oxygen and 3 ppm of water.) Gas of this purity may be used directly without further purification or drying. (Various absorption trains, etc., are often used in an attempt to dry or deoxygenate the gas further but generally have the opposite effect.)

SAFETY In particular, gas-washing (Dreschel) bottles containing concentrated sulphuric acid should NOT be used, as they are ineffective and can cause appalling injuries by splashing if suddenly overpressurized.

Argon and helium are also used but are much more expensive.

The gas cylinder should be equipped with a pressure-reducing 'diaphragm' regulator (0–25 psi) connected to a needle valve to control the gas flow rate. If a needle valve is not available, a screw clip on the gas line close to the valve is a good substitute.

SAFETY Diaphragm regulators are prone to 'stick' when turning the pressure up from zero and this can lead to a sudden pressure surge. The preferred technique is to close the needle valve (or screw clip), thus isolating the regulator from the reaction apparatus, adjust the regulator to the required supply pressure (3–5 psi), and then open the needle valve (or screw clip) to set up the required flow rate.

If nitrogen cyclinders are in short supply or cannot be moved to where the experiment is being carried out, then balloons filled with nitrogen can be used in some cases (e.g. Fig. 2.17). However, the nitrogen does not long remain in a high state of purity because of inward diffusion of oxygen and water vapour through the balloon membrane.

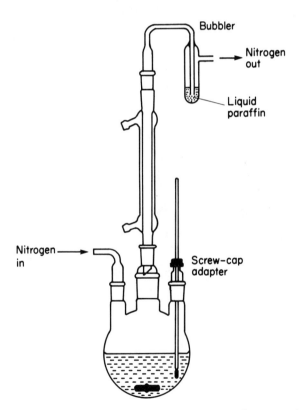

Fig. 2.15 Apparatus assemblies for reactions carried out under an inert (nitrogen) atmosphere.

(ii) Apparatus assemblies for reactions under inert atmospheres

As for anhydrous reactions ground-glass joints must be greased (sparingly) and a well sealed gland used for paddle stirrers. Simple modifications of standard preparative equipment are required to allow the system to be flushed through (purged) with nitrogen (or other inert gas) and then kept under a slight positive pressure throughout the reaction. This is usually achieved by some variation of the 'bubbler' technique (e.g. Fig. 2.15). The bubbler provides the exit path for the nitrogen and prevents air diffusing into the apparatus. Bubblers are of many types (Figs. 2.15 and 2.16) and are usually filled with medicinal liquid paraffin (less hazardous than mercury). The nitrogen inlet can be located in several places in the assembly. If well away from the bubbler (Fig. 2.15) the preliminary flushing out is facilitated and can be done without disassembly of any parts of the apparatus. A nitrogen inlet via a hypodermic needle piercing a rubber septum (Section 2.2.1) is convenient for small-scale reactions. An inlet adjacent to the bubbler can also be used (e.g. Fig. 2.16) but in

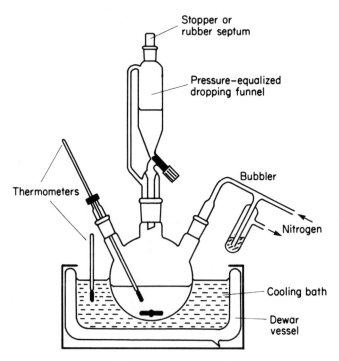

Fig. 2.16 Apparatus assembly for a reaction at low temperature and under an inert atmosphere.

such systems the stopper or rubber septum in the dropping funnel must be removed during flushing to allow a good flow of nitrogen through the apparatus. In such cases a combined bubbler/gas inlet, as shown, is convenient. For a reaction under reflux, a bubbler of this type would be fitted to the top of the condenser. Many variations on these basic themes are possible depending on the requirements of the experiment, and there is much scope for ingenuity in designing effective systems.

The nitrogen flushing operation may be carried out either before or after the flask is charged with solvent and reactant depending on the nature of the reaction. In elementary exercises the instructions will specify and in project work you will have to think it through. After flushing, the nitrogen flow is turned down to a very slow bubbling rate. However, do remember to turn up the flow rate when the apparatus is cooling down after a reaction, to avoid air being drawn in through the bubbler as the gases and vapours in the system contract on cooling.

(iii) Addition of reactants under inert atmospheres
It is assumed here that the reactants themselves are not highly reactive to air or water vapour (for those that are, see Section 2.2.1).

LIQUIDS AND SOLUTIONS. In moderate- or large-scale reactions these can be measured out rapidly in a measuring cylinder and dripped in via a pressure-equalized dropping funnel (e.g. as in Fig. 2.16). Small volumes (up to say 10 ml) are more easily handled using a syringe and injected into the reaction vessel via a rubber septum (the use of syringes and septa are discussed in detail in Section 2.2.1).

SOLIDS. In many cases solid reactants can be placed in the reaction vessel before flushing with nitrogen (e.g. magnesium in Grignard reactions). The gradual addition of solids during a reaction while maintaining an inert atmosphere is not easy and they should be added as solutions whenever possible. On a moderate/large scale, motor-driven auger-type delivery systems are very effective, but are rarely available outside industrial laboratories. The most common laboratory method is some variation on the system shown in Fig. 2.17 in which the solid is loaded into a bent tube and fitted to an angled neck on the reaction flask. Rotation and tapping allows gradual addition under reasonable control. The tap on the solids tube is optional but useful when flushing the system with nitrogen. Obviously this sytem cannot be used as shown when the reaction is being carried out under

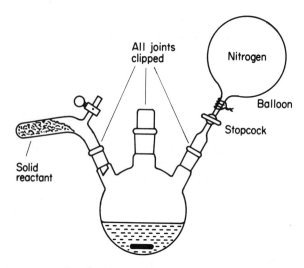

Fig. 2.17 Apparatus for the slow addition of a solid reactant and showing the use of a balloon for the supply of nitrogen.

reflux; then it is necessary to interpose a short wide-bore condenser between the flask and the tube.

2.2 SPECIAL TECHNIQUES

2.2.1 The use of air- and water-sensitive reagents

Changes and improvements in synthetic methodology in recent years have seen the rapidly increasing use of reagents that are very reactive to air and/or water, and it is important that all chemists are trained in the techniques for using them safely. In this section we deal with the basic guidelines for transferring reactive liquids and solids. Further information on these and other techniques is available from a number of excellent articles and technical bulletins[7-11].

7. *Handling Air-Sensitive Reagents*, Aldrich Technical Bulletin, AL-134.
8. *Equipment for Handling Air-Sensitive Reagents*, Aldrich Technical Bulletin, AL-135.
9. Gill, G. B. and Whiting, D. A. (1986) Guidelines for handling air-sensitive compounds, *Aldrichimica Acta,* **19** (2), 31.
10. Shriver, D. F. (1981) *The Manipulation of Air-Sensitive Compounds*, McGraw-Hill, New York.
11. Kramer, G. W., Levy, A. B., Midland, M. M. and Brown, H. C. (1975) *Organic Synthesis via Boranes*, Wiley, New York, ch. 9.

We thank Aldrich Chemical Co. Ltd. for permission to use material from references 7–9.

SAFETY There are well developed techniques by which air- and water-sensitive reagents can be used easily and safely. However, practitioners must always remember that many of these reagents are potentially very hazardous as they react violently with water, and some are pyrophoric (they inflame spontaneously on contact with air). Operators should ensure that they have taken adequate safety precautions (Section 1.3) bearing in mind that some of these operations involve pressurized vessels. In particular, the operations described in this section should NOT be performed by undergraduates or other inexperienced workers unless they are properly supervised throughout the operation by trained personnel. It is good training and sound practice to rehearse each experiment in full using an innocuous liquid or solid before attempting it with the reactive material.

(a) Liquid reagents and solutions

Inorganic liquids such as titanium tetrachloride and stannic chloride fall into this class together with organometallic species such as organo-lithiums, Grignard reagents and organo-boranes, which are all readily prepared as solutions in ether or hydrocarbon solvents. Many such reagents are now available commercially as standardized solutions in sealed packs (e.g. the Aldrich Sure/Seal® bottles). These reagents must not be exposed to the air in preparation, transfer or use.

Extensions to the normal procedures for anhydrous and inert-gas working (Section 2.1.5) are only required when these liquids have to be transferred from one vessel to another – for example, from the flask where the reagent has been prepared to the dropping funnel of another piece of apparatus. Similar techniques are required for the transfer of the commercially available reagents (such as solutions of organo-lithium reagents) from their supply bottles to either a reaction vessel or a dropping funnel.

These liquid transfers can be easily accomplished by the needle/septum methodology outlined below. On a small scale this involves the use of a syringe, and on a larger scale the use of a double-ended transfer needle. Both techniques require the use of rubber septa (Fig. 2.18) on storage bottles (e.g. Fig. 2.21), reaction vessels (e.g. Fig. 2.22) and dropping funnels (e.g. Fig. 2.23). These septa are easily pierced by hypodermic needles and reseal when the needle is withdrawn. Septa (e.g. 'Suba-Seal®' rubber septa) can be obtained to fit most socket and tubing sizes. They should be wired on (Fig. 2.18b) if positive pressure is used (see below) or if the septum will come into

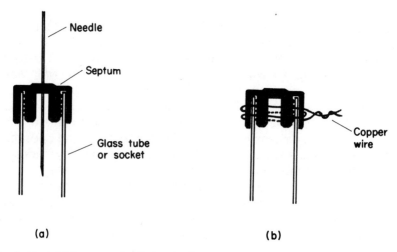

Fig. 2.18 (a) The use of a 'Suba-Seal' septum; (b) securing the septum with thin copper wire.

contract with solvent vapour since this can cause swelling and ejection of the septum from its socket. For reactions under reflux it is better to minimize the contact area by using an adapter (Fig. 2.19) and a small (*c*. 6 mm) septum, rather than using a large septum to fit the socket. The lifetime of septa can be maximized by using a little silicone grease on the top of the septum to ease the passage of the needle.

Both the syringe and double-needle transfer methods require the provision of a supply of dry nitrogen at low pressure (1–3 psi) (Section 2.1.5) via a flexible line (plastic tubing) terminating in a short 18 or 20 gauge hypodermic needle. A nitrogen bubbler equipped with a rubber septum is also required so that syringes can be flushed with nitrogen by filling and emptying. An arrangement such as that shown

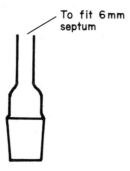

Fig. 2.19 Septum adapter.

in Fig. 2.20 serves both purposes. The ability to control the pressure in the 0–5 psi range is important and a good regulator and pressure gauge should be incorporated.

(i) Syringe techniques

Syringes are usually of the all-glass type in capacity up to 50 ml. The needles need to be long (12–24 inches) and flexible, normally of 18 or 20 gauge but 16 gauge may be required for viscous liquids. Use of the syringe is safer and easier if a Luer-lock syringe stopcock is fitted between the syringe and the needle (Fig. 2.21) but it is not essential. All parts should be dried and cooled separately (Section 2.1.5) and assembled using a touch of silicone grease on the Luer joint (and the stopcock if fitted) to ensure gas-tightness.

Syringes may be used for the measurement and rapid transfer of small quantities of reagent (up to *c*. 50 ml). They should not generally be used for the slow addition of the reagent over a long period since some hydrolysis inevitably occurs around the plunger and this can lead to seizure if the plunger is left static for long. Seizure can be prevented by using a drop or two of silicone fluid or mineral oil to lubricate the plunger (if the contamination is acceptable). However, it is generally better to transfer the reagent to a pressure-equalized dropping funnel if slow addition is required.

A reagent may be transferred from a storage bottle fitted with a rubber septum (e.g. Aldrich Sure/Seal® bottles) in the following way (see Safety note on p. 33). The bottle must be firmly located using a retort-stand ring (Fig. 2.21). Using the nitrogen supply line (Fig. 2.20) set the source pressure at 1–2 psi and adjust the flow controller to give a slow flow through the bubbler (tap A open). Then insert the short needle B through the septum of the bottle, wait a moment and then close tap A. This procedure will first vent any excess pressure that may have built up in the bottle during storage and then slightly

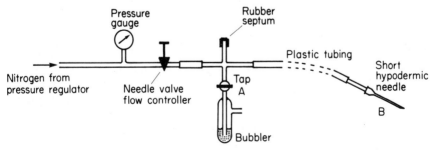

Fig. 2.20 Nitrogen supply line.

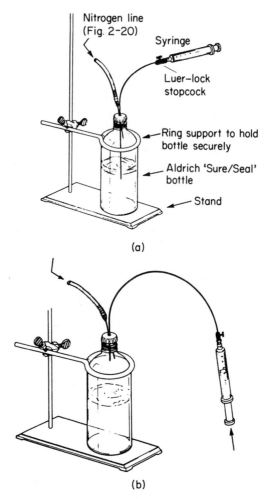

Fig. 2.21 (a) Filling the syringe from the reagent bottle (Aldrich Sure/Seal bottle); (b) removing gas bubbles and returning excess reagent to the bottle.

pressurize the storage bottle to 1–2 psi (2 psi is normally the maximum needed for this operation – do NOT overpressurize). Close the syringe Luer-lock stopcock and insert the syringe needle through the septum and into the liquid (Fig. 2.21a). Open the syringe stopcock; the positive pressure will drive liquid into the syringe so you must control the plunger carefully as it is forced back. It can be eased back gently if it sticks but do not pull it back strongly as this causes leaks and creates gas bubbles in the syringe. Take in slightly more reagent than required and then move the syringe to the position shown in Fig.

2.21b so that any gas bubbles are at the 'needle' end. Then depressurize the bottle by opening tap A (Fig. 2.20), and expel any gas and excess reagent from the syringe back into the bottle. When excess reagent has been expelled, draw a little nitrogen into the syringe to clear the reagent from the needle and then close the syringe stopcock before withdrawing the needle from the bottle. Transfer quickly and inject into the reaction apparatus through a septum.

Solutions may be transferred from reaction flasks equipped with septa in a similar way. In such assemblies (e.g. Fig. 2.22) the septum should be wired and all cone/socket joints clipped together with a joint clip (Fig. 2.1).

(ii) Double-ended needle transfers

Volumes of solution larger than *c.* 50 ml are more conveniently transferred by this method. Transfer is carried out via a long double-tipped

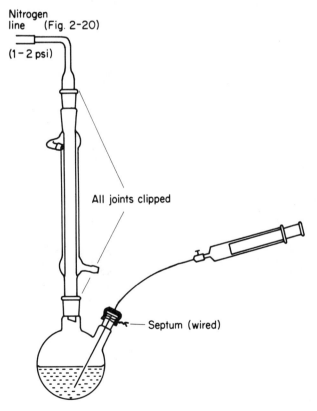

Fig. 2.22 Filling the syringe from a reaction flask.

needle (available in 24 or 36 inch lengths of 16–20 gauge stain-
less steel) or as two wide-bore needles (12 gauge) 18 and 6 inches
long connected by a 30 inch length of Teflon® tubing (Fig. 2.23)
(Aldrich Flex-needle®). The Flex-needle® is ideal for transfers to and
from vessels fitted with Suba-Seal® rubber septa as they will accept
the wide needles and will reseal again afterwards. However, for
transfers from Aldrich Sure/Seal® reagent bottles it is recommended
that the narrower all-steel double-ended needles (maximum 16
gauge) are used since the septa fitted to these bottles do not reseal
well after puncture by the 12 gauge Flex-needle®.

Transfer from a reaction flask via a Flex-needle® may be carried
out by the following method (see Safety note on p. 33). It is important
that all cone/socket joints in the apparatus are secured with joint
clips (Fig. 2.1) and septa are wired on. First, set up the nitrogen line
(Fig. 2.20) to 1 psi as described above and connect it to the vessel
either via a cone/tubing adapter as shown in Fig. 2.23 or via a needle/
septum connection. Then close tap A (Fig. 2.20) to pressurize the

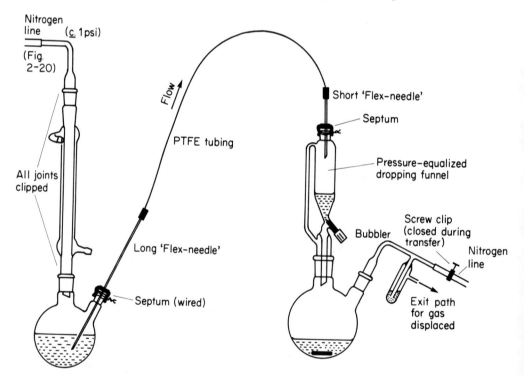

Fig. 2.23 Use of the Aldrich Flex-needle for transferring a solution from a
flask to a dropping funnel under inert-atmosphere conditions.

flask to *c.* 1 psi (normally the maximum required). Insert the long needle of the Flex-needle® just through the flask septum and allow nitrogen to flow through the transfer line to flush it out; then insert the short needle through the septum of the receiving vessel (Fig. 2.23). Push the long needle down into the liquid in the flask to effect the transfer. When enough has been transferred, raise the long needle again – this stops the flow and blows out any liquid in the transfer line.

A similar technique is used to transfer reagents from Aldrich Sure/Seal® reagent bottles but using a narrower needle as mentioned above and a needle/septum nitrogen connection to pressurize the bottle as in Fig. 2.21. Again only low pressure (1–2 psi) is required.

A problem in double-ended needle transfers is measuring the volume of reagent. If the transfer is to a dropping funnel as in Fig. 2.23 then the simplest way is to calibrate it by taping a pre-measured paper scale onto the outside. Otherwise the reagent must first be transferred to a measuring cylinder (Fig. 2.24a) using an adapter fitted with two septa (Aldrich) and then to the reaction apparatus (Fig. 2.24b).

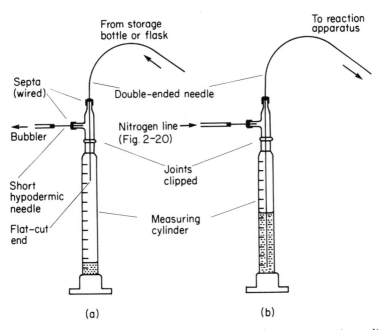

(a)   (b)

Fig. 2.24 The transfer of a solution (a) to and (b) from a measuring cylinder under inert-atmosphere conditions. All joints should be clipped and septa should be wired on.

Guidance should be obtained from your instructor on washing up syringes and transfer lines and on the disposal of any excess reagent.

(b) Solid reagents

Common reagents in this class are hydrides (e.g. lithium aluminium hydride and sodium hydride), alkoxides (e.g. sodium methoxide and potassium t-butoxide), aluminium trichloride and phosphorus pentoxide. All of these are reactive to water vapour and should be weighed out and transferred under protection.

SAFETY Lithium aluminium hydride is particularly hazardous because it reacts violently with water, alcohols and other protic solvents, releasing hydrogen, which usually inflames spontaneously. It is a remarkably useful reagent but it should be used by inexperienced workers ONLY under the strictest supervision.

Absolutely dry conditions can only be provided by a good glove box but this degree of protection is rarely necessary in preparative work. A plastic glove bag (e.g. the Aldrich 'Atmosbag®' (Fig. 2.25)) is usually adequate for weighing out material and transferring it to the reaction apparatus. Glove bags are made of clear polythene film with a large opening for loading and smaller ones for gas, electricity and vacuum lines. After loading all the equipment required – reaction apparatus, balance, weighing bottles, funnel, spatula, etc. – the large opening is taped up with removable tape. The bag is then purged of air by repeated evacuation – which collapses it around the apparatus, so avoid sharp edges – and filling with dry nitrogen. Three cycles are usually enough and the bag is then inflated, and the operation carried out.

Afterwards any unused portion of the reagent must be adequately sealed up before the bag is opened. This is particularly important with compounds such as lithium aluminium hydride, which is supplied in sealed polythene bags contained in tins that cannot be resealed. A dried sealable container for the unused portion must therefore be included when loading the bag.

As with operations involving reactive liquids, these techniques should be rehearsed in every step before attempting them with the air-sensitive solid reagent. When carrying out manipulations within the dry bag the operator cannot easily get out of the gloves so it is highly desirable to have a colleague at hand in case the inert-gas supply needs to be adjusted or any other external operation has to be done.

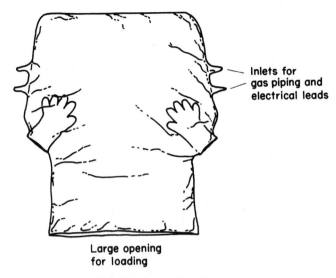

Inlets for
gas piping and
electrical leads

Large opening
for loading

Fig. 2.25 A 'dry bag' (Aldrich 'Atmosbag').

SAFETY In particular, the operator should consider personal protection in the event that the bag should be torn or punctured. It is essential to wear double gloves (thin surgical gloves give extra protection with the minimum loss of dexterity), and to set up the dry bag in a fume-cupboard or hood.

2.2.2 Reactions in liquid ammonia

Liquid ammonia (b.p. $-33°C$) is a useful solvent particularly in reactions requiring the use of the alkali-metal amides (e.g. sodamide, $NaNH_2$) where it serves both as solvent and reactant in the formation of the base.

SAFETY It is quite easy to use liquid ammonia in the laboratory but, since it is extremely noxious and toxic, it is vital that undergraduates and other inexperienced workers use it ONLY under the strictest supervision by qualified personnel. It must be used only in a very efficient fume-cupboard in an area where all personnel can be rapidly evacuated in the event of an accident, or if the fume-cupboard extraction system should fail. Adequate personal protection (Section 1.3) should be worn.

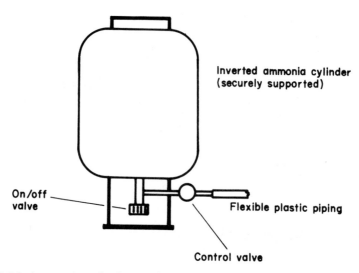

Fig. 2.26 Ammonia cylinder used in an inverted position to provide liquid ammonia.

Liquid ammonia is supplied in cylinders of various sizes, the small (6 kg) ones being of convenient size and shape for laboratory use. The ammonia is obtained in liquid form by using the cylinder in an inverted position – supported on the integral stand – with the valve at the bottom (Fig. 2.26). Whatever size of cylinder is used, it must be properly supported and secured. Before use the cylinder should be equipped with the appropriate control valve (in addition to the main on/off valve on the cylinder) and a length of flexible delivery tubing wired onto the valve.

(a) Reaction assemblies

Reactions can be carried out in normal three-necked flasks (Fig. 2.27). If the reaction is to be done at the b.p. of ammonia ($-33°C$) the flask should be supported on a cork ring inside a large polythene bowl or saucepan packed loosely with insulation material such as vermiculite. This arrangement also provides containment in case the flask should be broken. For reactions at lower temperatures the flask should be placed in a coolant bath in the usual way (Section 2.1.4). It is essential to fit a Cardice reflux condenser (Fig. 2.27) to minimize the loss of ammonia. The flask can be equipped with a paddle or magnetic stirrer, or a pressure-equalized dropping funnel, etc., in the usual way. The entry of moisture should be prevented by using guard tubes

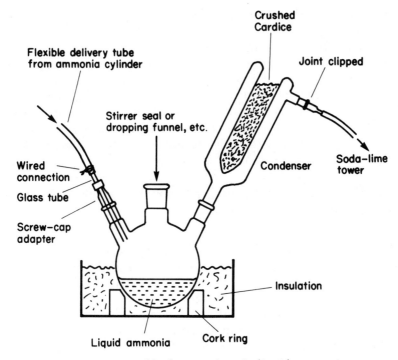

Fig. 2.27 Apparatus assembly for reactions in liquid ammonia.

packed loosely with granulated soda-lime (not calcium chloride), or a guard tower (Fig. 2.27), which has less resistance to gas flow.

(b) Charging the reaction vessel

When charging the reaction flask with liquid ammonia both the cylinder and flask should be in the fume-cupboard. The plastic delivery tube should be wired on at both ends.

Assemble the apparatus (e.g. Fig. 2.27) and place crushed Cardice (Section 2.1.4) in the well of the cold-finger condenser. Make sure that the fume-cupboard doors are pulled well down and that there is a good draught. Check that the cylinder control valve is off, open the main on/off valve and then gently open the control valve. The liquid ammonia will volatilize rapidly in the early stages until the delivery tube and flask have cooled down, so proceed cautiously. When the flask has been charged, turn off the control valve and the on/off valve, wait a few moments until the liquid has volatilized out of the transfer line and then detach it from the flask and replace the screw-cap adapter fitting with a stopper or pressure-equalized dropping funnel as required.

When adding the reactants bear in mind that many reactions are exothermic and hence rapid addition can cause the ammonia to boil vigorously. Therefore it is sensible to proceed with caution and to keep the Cardice condenser well topped up. You will find that a film of ice will form on the outside of the flask, making it difficult to see what is going on inside. When necessary you can clear a 'window' by scraping or with a squirt from an acetone or ethanol wash bottle and a wipe with a tissue to remove the solvent.

SAFETY Remember that these are flammable, volatile solvents and take appropriate precautions.

After the reaction has been carried out it is usual to leave the reaction mixture overnight or until the Cardice and liquid ammonia have completely evaporated. It is unacceptable to vent large quantities of noxious material through fume-cupboards so you should connect the exit from the guard tower to a suitable absorption device (consult your instructor) during this operation. (Carry out the work-up operations in the fume-cupboard.)

2.2.3 Catalytic hydrogenation

This process involves the combination of an organic reactant with gaseous hydrogen in the presence of a metal or metal oxide catalyst. The reaction is carried out by vigorously stirring or shaking a solution of the reactant containing the catalyst as a suspended powder, under an atmosphere of hydrogen (e.g. Fig. 2.28) until the uptake of hydrogen ceases. The pressure of hydrogen required and the reaction temperature vary with the nature of the functional group being reduced. Catalysts and reaction conditions are discussed extensively in references 12–14. These conditions fall into three categories and are carried out in different kinds of apparatus: (a) hydrogen at atmospheric pressure/room temperature; (b) hydrogen at low pressure (up to c. 60 psi) and (c) hydrogen at high pressure (above c. 60 psi). The

12. Augustine, R. L. (1965) *Catalytic Hydrogenation*, Edward Arnold, London/ Marcel Dekker, New York.
13. Freifelder, M. (1971) *Practical Catalytic Hydrogenation*, Wiley-Interscience, New York.
14. Rylander, P. N. (1979) *Catalytic Hydrogenation in Organic Synthesis*, Academic Press, New York.

procedure for (a) is described below but (b) and (c) are beyond the scope of this book.

SAFETY The potential hazards in hydrogenation are (i) hydrogen is highly inflammable and hydrogen/air mixtures are explosive, (ii) high pressures are used in some cases, and (iii) the catalysts used are highly reactive and can cause spontaneous ignition of solvent vapours both before and after use. Hydrogenation reactions should therefore be carried out by undergraduates and other inexperienced workers ONLY under the strictest supervision by qualified personnel.

(a) Atmospheric-pressure hydrogenator

The apparatus (Fig. 2.28) consists of a gas manifold from which connections lead to (i) a hydrogen cylinder and a mercury bubbler, (ii) several calibrated gas burettes of different sizes (one only shown) from which hydrogen is drawn during the reaction, (iii) a flexible connection to the reaction vessel, and (iv) a vacuum connection to a water (filter) pump.

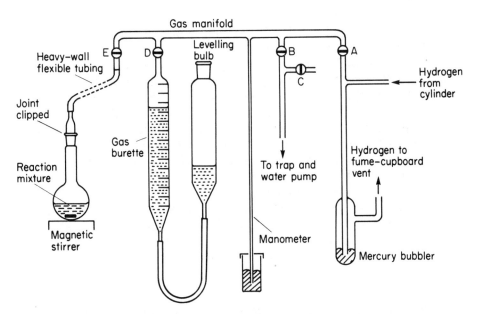

Fig. 2.28 Apparatus for hydrogenation at atmospheric pressure.

SAFETY The whole apparatus should be operated in a well vented fume-cupboard (with a spark-sealed extraction fan) and obviously there must be no flames, heaters, ovens, infra-red spectrometers or other sources of ignition in the vicinity. The vent pipe from the mercury bubbler should be located close to the extraction duct.

The overall operation involves first removing all air from the system by repeatedly evacuating the manifold and reaction vessel and filling with hydrogen. Then a gas burette of appropriate size is filled with hydrogen and the reaction solution is stirred vigorously until hydrogen absorption is complete. The detailed procedure is as follows*.

(i) Preparation of the reaction mixture
Weigh out the required amount of catalyst, place it in the hydrogenation flask and then add a solution of the reactant in an inert solvent (e.g. acetic acid, ethanol, methanol, cyclohexane or hexane) and a magnetic stirrer bar.

SAFETY The catalyst MUST be put into the flask BEFORE the solvent; the reverse procedure can cause ignition of the solvent.

The flask should not be more than half full. Attach the flask to the hydrogenator using a little silicone grease on the joint and secure it with springs or a joint clip (Fig. 2.1). Put a safety screen round the flask. Close tap E.

(ii) Removal of air from the system and filling the gas burette
1. With taps A and E closed, open taps B, C and D, and raise the levelling bulb to expel all gas from the gas burette; then close tap D. Put the levelling bulb at its lowest position.
2. Turn on the water pump, open tap E and then close tap C to begin the evacuation – the mercury level in the manometer will change as the system is evacuated.
3. While this is happening seek assistance from your instructor and set the flow control on the hydrogen cylinder to produce a brisk flow of hydrogen through the mercury bubbler and out of the vent tube.

---

* Hydrogenators will vary slightly in design but the operating method for each should be easy to work out from the description given here once the principle is understood.

4. When the mercury level in the manometer is steady, close tap B and open tap A gently to fill the manifold and reaction vessel with hydrogen.
5. Close tap A and open tap B to evacuate the system again.
6. Repeat step 4, step 5 and then step 4 again so that at this point the manifold and reaction vessel are filled with hydrogen, tap B is closed and tap A is open.
7. Open tap D to admit hydrogen to the gas burette. When it is full, close tap A and turn off the hydrogen supply.
8. Open tap C and turn off the water pump.

(iii) The hydrogenation

Equalize the liquid levels in the gas burette and record the level. Turn on the magnetic stirrer to agitate the reaction mixture vigorously and follow the hydrogen absorption by occasionally adjusting the levelling bulb. When the uptake ceases, equalize the levels and record the final burette reading. Do a calculation to see if the volume absorbed accords with what was expected. In some cases the reaction can come to a premature end because of catalyst poisoning and it may be necessary to add more catalyst.

SAFETY The catalyst MUST be added as a suspension in the reaction solvent NOT as a dry powder (see (i) above).

(iv) Disassembly and work-up

1. Close tap D and turn off the magnetic stirrer to let the catalyst settle. Then swirl the flask to wash down any catalyst adhering to the sides of the flask but do NOT remove the hydrogenation flask at this stage.
2. Turn on the water pump, close tap C and open tap B to evacuate the gas manifold and reaction flask.
3. When the manometer is steady, open tap C and then tap B (cautiously) to admit air to the system.
4. Remove the hydrogenation flask and work up the mixture as described in step 5.

SAFETY Inexperienced workers should be supervised during this operation since catalysts that have been exposed to hydrogen can ignite spontaneously if they are allowed to dry out.

---

5. Filter the reaction mixture to remove the catalyst using a sintered funnel with the sinter covered by a thin layer of Celite, using light suction, but do NOT let the catalyst dry. Wash the catalyst with a little solvent and again do not let it dry. Remove the filtrate for work up and immediately wash the catalyst thoroughly with water and consult your instructor about its disposal. Wash the hydrogenation flask and magnetic stirrer thoroughly with water to remove traces of catalyst.
6. Carry out step 1 in subsection (ii) to prepare the apparatus for the next user, and turn off the water pump.

2.2.4 Photochemistry

(a) Reaction apparatus

Preparative photochemical reactions are usually carried out in reactors of the type shown in Fig. 2.29*. In these the ultraviolet (UV) lamp is contained in the central well, surrounded by a cooling jacket, immersed in the solution to be irradiated. Various sizes of reaction vessel can be used so that the lamp can be completely immersed to maximize the absorption of the radiation. The immersion well can be made from quartz (very expensive), which transmits radiation down to c. 200 nm, or more cheaply from Pyrex® (or other borosilicate glass), which is transparent down to only c. 300 nm. The reaction vessels are made of Pyrex®. The cooling jacket around the lamp is essential when using the popular medium-pressure mercury lamps (see below) that run at high temperature and allows the reaction solution to be kept at room temperature (or at other selected temperatures by immersing the reaction vessel itself in a thermostat bath and circulating water at the same temperature through the jacket). It is usually necessary to stir the solution during irradiation and this can be done using a magnetic stirrer or by bubbling a stream of nitrogen or other inert gas through the solution.

SAFETY Ultraviolet radiation is very damaging to the eyes and skin. Reactors must therefore be completely screened while in operation. If it is necessary to work on the equipment while the lamp is lit, then the operator must wear special eye-protection goggles (absorbing all radiation below 400 nm) even for the shortest exposure and also have skin protection by clothing, gloves and high-protection sun-tan cream.

---

* Reactors of this type are supplied by Applied Photophysics Ltd, 203–205 Kingston Road, Leatherhead, KT22 7PB.

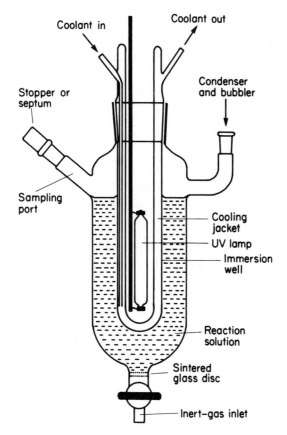

Fig. 2.29 A typical immersion-well photochemical reactor.

(b) Ultraviolet source lamps

The UV sources most often used are mercury vapour lamps (details of other types of lamps are given in reference 15).

(i) Low-pressure mercury lamps

These are of low power (5–20 W) and emit *c*. 90% of their radiation at 254 nm (the other main emission band at 185 nm is not transmitted by quartz). Further details are given in reference 16. They are cool running and need minimal coolant flow through the jacket when used in an immersion-well reaction (Fig. 2.29).

15. Schenk, G. O. (1968) in *Preparative Organic Photochemistry* (ed. G. Schönberg), Springer-Verlag, Berlin, ch. 46.

(ii) Medium-pressure mercury lamps

These are of much higher power (100–400 W) and take *c.* 15 min to reach their high working temperature and full output. They give out much heat and are generally used in immersion-well reactors (Fig. 2.29) to avoid heating the reaction solution. They emit radiation over a wide range with a very strong UV band at 365–366 nm, weaker UV bands at 334, 313, 303, 297, 265 and 254 nm and also much visible radiation at 404–408, 436, 546 and 577–579 nm. See reference 16 for further details.

This wide range of radiation makes medium-pressure lamps highly effective for bringing about photochemical reactions with a wide range of organic compounds having different UV absorption spectra.

If irradiation at a selected wavelength is required, then the other wavelengths can be filtered out using various types of glass[17] or solutions of inorganic salts for which details can be found in Table 2.3 and the reference given therein.

(c) Reaction conditions and reaction solvents

Photochemical reactions are usually carried out using fairly dilute solutions (*c.* 0.01–0.05 M) and with good stirring. High dilution

Table 2.3 Cut-off filter solutions[a]

Cut-off wavelength (nm)	Composition
below 250	Na_2WO_4
below 305	$SnCl_2$ in aq. HCl (0.1 M in 2:3 HCl/H_2O)
below 330	2 M Na_3VO_4
below 355	$BiCl_3$ in HCl
below 400	KH phthalate + KNO_2 (in glycol at pH 11)
below 460	K_2CrO_4 (O.1 M in NH_4OH/NH_4Cl at pH 10)
above 360	1 M $NiSO_4$ + 1 M $CuSO_4$ (in 5% H_2SO_4)
above 450	$CoSO_4$ + $CuSO_4$

[a] For more details and other solutions see Zimmerman, H. E. (1971) *Mol. Photochem.*, **3**, 281. Reprinted by courtesy of Marcel Dekker, Inc. This paper gives more details on these and other filter solutions.

16. Calvert, J. G. and Pitts, J. N., Jr (1966) *Photochemistry*, Wiley, New York.

allows good penetration of the radiation into the reaction solution and helps to minimize the deposition of polymers and tars on the wall of the immersion well around the lamp, where they can block off the light and stop the reaction.

Many types of solvent can be used, ranging in polarity from hydrocarbons to alcohols.[17] Obviously the solvent should not significantly absorb radiation at the frequency required to excite the reactant[18] and it should not itself undergo photo-decomposition. In some cases it may also be important that the solvent is inert to any reactive intermediate generated in the photolysis. All solvents should be carefully purified to remove any peroxides and/or UV absorbing impurities (see references on pp. 24 and 91), and then checked by running a UV spectrum. As a matter of routine it is sensible to run a UV spectrum of new batches of solvent to check for UV-absorbing impurities.

Dissolved oxygen can have deleterious effects on many photochemical processes and the reaction solution should be deoxygenated in the reaction vessel before irradiation. This is most easily done by bubbling through it a stream of oxygen-free nitrogen, helium or argon for *c.* 30 min. Obviously the reaction should then be carried out under inert-atmosphere procedures (Section 2.1.5).

2.2.5 Flash vacuum pyrolysis

Pyrolysis reactions are those in which the reactant is converted into products by heat alone. However, pyrolysis in solution often results in the formation of multiple products, polymers and tars due to bimolecular reactions of the reactant (or reactive intermediates formed) with itself and/or the solvent. These problems are avoided in flash vacuum pyrolysis (FVP) in which the reactant molecules are decomposed in the vapour phase, under high vacuum, by a very short exposure to high temperature. Under these conditions the reactant molecules are well separated in the gas phase and so bimolecular processes are kept to a minimum and there is no solvent to react with the hot molecules or with reactive intermediates formed in the thermolysis.

A typical FVP apparatus is shown in Fig. 2.30. The pyrolysis tube is made of quartz and is *c.* 30 cm long by 2.5 cm diameter. It is heated by a furnace capable of accurate temperature control in the range

17. Horspool, W. M. (ed.) (1984) *Synthetic Organic Photochemistry*, Plenum Press, New York.
18. *UV Atlas of Organic Compounds*, (1965) Butterworths, London.

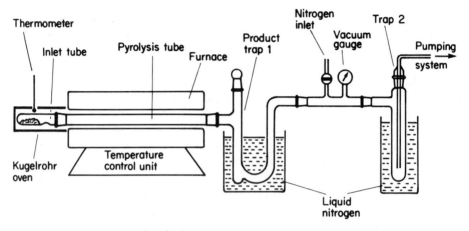

Fig. 2.30 Apparatus for flash vacuum pyrolysis.

200–1000°C (e.g. a Stanton–Redcroft LM8100 tube furnace). The reactant is vaporized into the pyrolysis tube from an inlet system (most conveniently a tube heated by a Buchi Kugelrohr oven), and the products are collected in traps cooled by liquid nitrogen. The system is evacuated to 10^{-1} torr or better during pyrolysis. Most FVP experiments are carried out at pressures in the 10^{-1}–10^{-2} torr range, which can be achieved easily with a rotary oil pump. However, there are a few that need lower pressures and hence the use of an oil diffusion pump, but these are rare.

Pyrolysis experiments are easy to carry out. Before starting the actual experiment clean up the pyrolysis tube by heating it in air (to c. 800°C), with the inlet tube and traps disconnected. Cool it down, place the reactant in the inlet tube and assemble the traps as shown in Fig. 2.30. Charge trap 2 with liquid nitrogen and switch on the pump to evacuate the system to the required pressure (see below). Switch on the furnace and allow it to equilibrate at the required temperature (see below). Any volatile contaminants in the system or the reactant will condense in trap 2. Then charge the product trap (trap 1) with liquid nitrogen and set the inlet heater at a temperature that will vaporize the reactant into the pyrolysis tube at c. 0.5–1.0 g h^{-1}. When all the reactant has passed through, turn off the furnace and let it cool, turn off the vacuum pump, admit dry nitrogen to the system and allow the traps to warm up to room temperature. The product can then be washed out of the trap with a solvent (e.g. dichloromethane) and worked up in the usual way.

SAFETY Take normal precautions for vacuum work (Section 1.3.3). Before starting the pyrolysis experiment consider the chemistry carefully and work out how you are going to deal with any toxic volatile material (e.g. HCN, hydrogen halides, SO_2, NH_3) that may be extruded in the pyrolysis and condensed in the cold trap together with the required organic product. It may suffice to vent the traps into a suitable chemical absorption train as they warm up. Alternatively the volatile material can be separated by bulb-to-bulb distillation, i.e. while the system is still under vacuum keep trap 2 cooled in liquid nitrogen and allow trap 1 to warm up sufficiently for the volatile material to distil over (see Section 2.1.4, for coolant baths).

There are now many reviews on FVP and its use in preparative work[19-21] that provide information about the scope of the technique and about the temperatures and pressures used for the pyrolysis of various classes of compound. The very high furnace temperatures used (400–1000°C) may at first sight give the impression that this is a sledgehammer technique of undiscriminating brutality. However, this is not so; the combination of high temperature with very short contact time (typically 1–20 ms) gives FVP an unexpected delicacy and selectivity in bond cleavage. Temperature changes of as little as 50°C can change the reaction path followed and most FVP reactions go very cleanly in high or quantitative yield.

The above describes only the basic preparative use of FVP. In more sophisticated applications it can be used to prepare highly unstable short-lived materials, and also in mechanistic work in which the pyrolysis products are trapped at low temperature for direct examination by infra-red and other spectroscopic techniques.

19. Wiersum, U.E. (1982) Flash vacuum thermolysis, a versatile method in organic chemistry. Part I. General aspects and techniques, *Recl. Trav. Chim. Pays-Bas*, **101**, 317–32.
20. Wiersum, U. E. (1984) Preparative flash-vacuum thermolysis. The revival of pyrolytic synthesis, *Aldrichimica Acta*, **17**, 31–40.
21. Brown, R. F. C. (1980) *Pyrolytic Methods in Organic Chemistry*, Academic Press, New York.

3 Isolation and purification of reaction products

3.1 PRIMARY WORK-UP PROCEDURES

3.1.1 General considerations

Preparative organic reactions rarely give a complete conversion of the starting material into the required product. In a good 'synthetic' reaction a yield of product of 70–80% or more would be expected, but in addition there may be small amounts of other organic materials formed as by-products. In some cases reactions also produce polymeric 'tarry' material, which may be a brown or yellow colour. In addition to these organic products, many reactions also produce equimolar amounts of inorganic products, e.g. the metal halides produced in a typical ether synthesis:

$$RCH_2Br + NaOCH_3 \rightarrow RCH_2OCH_3 + NaBr$$

Thus the mixture obtained after a reaction has been carried out can be quite complex even for a good synthetic reaction. It will consist of the reaction solvent, the major product, by-products, unreacted starting material, possibly polymeric material and possibly an inorganic product. Obviously the objective is to separate out the major product in as high a yield as possible and in as pure a state as possible. There are some reactions (unfortunately very few) in which the major product simply crystallizes out when you cool the reaction mixture down to room temperature. However, the procedure required is usually more complex and involves the successive application of several different techniques. This is generally called 'working up' the reaction mixture. The particular methods required will depend on the nature of the mixture and on the chemical and physical properties of the compounds concerned. In straightforward reactions with one major product the sequence of operations is usually designed to remove the reaction solvent and the inorganic products first, and then to separate the major organic product from the other organic materials present by crystallization (for solids) or distillation (for liquids). In more complex reactions there may be several 'major' products and it may be necessary to separate them by some form of chromatography (Chapter 4).

In most 'teaching' experiments and in books on synthetic chemistry the work-up procedure is usually specified in some detail, and even in the descriptions of new experimental work in the chemical journals it is usual to find a description of how the reaction was worked up and the products isolated. Therefore it is rare for those under training to have to decide how to work up a reaction, but it is important at all stages to make sure you understand why a particular operation is being carried out and what it is meant to achieve. This will prepare you for the time when you may be doing project work where the chemistry will be new, and for each reaction you will have to work out the best way to get the products separated and purified. In this you need to call on the wide range of practical skills covered in the following sections, and to use your experience to decide which ones are likely to be most effective for the reaction mixture in question.

3.1.2 Removal of solvent by rotary evaporator

At some stage in the work-up procedure it will be necessary to remove the reaction solvent – or the solvent used in an extraction procedure (Section 3.1.3). The instructions may simply say 'evaporate' the solvent. This does NOT mean that it should be boiled off into the atmosphere from an evaporating basin. Solvents are removed either by distillation (Section 3.4) or, in the vast majority of cases, by the use of a rotary evaporator (Fig. 3.1).

The rotary evaporator is essentially a device for the rapid removal of solvents by distillation under vacuum. Its use requires the following sequence of operations. Place the solution to be evaporated in a round-bottomed (RB) flask of appropriate size (not more than half full). Attach it to the vapour duct, using an adapter if necessary, and put a clip on the joint to make sure the flask does not fall off into the water bath. The joint need not be greased when using water-pump vacuum. Make sure the receiver flask is empty and is also clipped on. At this stage the water bath should be cold. Lower the apparatus so that the flask is *c.* one-third immersed in the water, turn on the condenser water, and then turn on the water-pump vacuum system (Section 3.4.8). As the vacuum develops, turn on the motor so that the

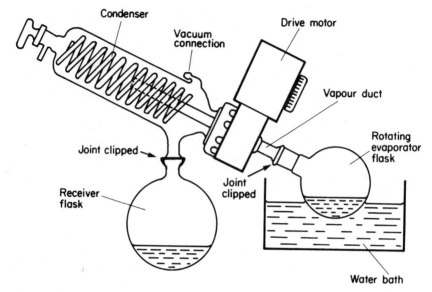

Fig. 3.1 Rotary evaporator.

evaporator flask rotates. When the vacuum has stabilized* start to heat the water bath slowly until the solvent begins to distil. Do not overheat or the liquid may foam up into the vapour duct.

The sequence is important: NEVER plunge the evaporator flask into a HOT water bath; it will most likely produce vigorous boiling and foaming and loss of material into the receiver.

Use of the evaporator with water-pump vacuum will allow the easy removal of 'volatile' solvents of b.p. up to *c*. 100°C, using a bath temperature of up to 50–60°C. For higher-boiling solvents it is preferable to use the high-vacuum version (Fig. 3.2).

(a) The high-vacuum rotary evaporator

In this modification the water condenser is replaced by a cold trap, cooled with liquid nitrogen or Cardice/acetone (Section 2.1.4). The vacuum is provided by a rotary oil pump (Section 3.4.8). The method of use is similar to the above but it is essential to make sure that the

---

* A good vacuum is essential. If *c*. 20–30 mmHg or better cannot be obtained then consult your instructor. Follow appropriate safety procedures for vacuum work.

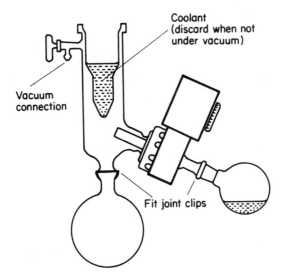

Vacuum
connection

Coolant
(discard when not
under vacuum)

Fit joint clips

Fig. 3.2 'High-vacuum' rotary evaporator.

liquid to be evaporated does not contain any 'volatile' solvents. If it does, then the liquid may foam vigorously when the vacuum is applied, even at room temperature. It is generally safer to use the normal rotary evaporator first, heating the bath up to 60°C or so, to remove the volatiles, and then to cool the liquid back down to room temperature before using the high-vacuum evaporator.

3.1.3 Extraction procedures

(a) The use of the separating funnel

The most straightforward extractions are carried out in a separating funnel (Fig. 3.3). The procedure involves mixing an organic solution (in a solvent that is not miscible with water) with water (or aqueous acid or alkali), shaking the funnel to mix the two layers thoroughly, allowing the mixture to stand until the two layers have separated again, and then running off the lower layer into another vessel. The mixing allows material to pass from one layer to the other depending on its relative solubility in each.

Basically extractions are of two kinds: (i) extraction with water to remove water-soluble material (usually inorganic) from a mixture, and (ii) extraction with aqueous acid or aqueous base to remove, respectively, organic bases (equation (3.1)) or organic acids (equation (3.2)) from the organic layer:

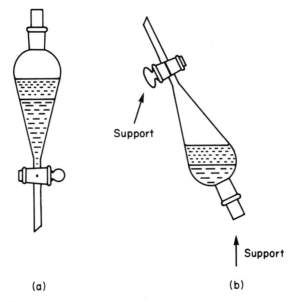

Fig. 3.3 (a) Separating funnel; (b) position for venting excess pressure.

$$RNH_2 \quad + \quad HCl(aq.) \quad \rightarrow \quad R\overset{+}{N}H_3Cl^- \quad (3.1)$$
$$\text{(water-insoluble)} \qquad\qquad\qquad \text{(water-soluble)}$$

$$RCO_2H \quad + \quad NaOH(aq.) \rightarrow \quad RCO_2^-Na^+ \quad (3.2)$$
$$\text{(slightly water-soluble)} \qquad\qquad \text{(very water-soluble)}$$

PROCEDURE. Select a separating funnel of sufficient volume that it will not be more than two-thirds full. Check the tap to make sure it rotates freely (use a little tap grease at the outer edges of glass taps, but not with Teflon® ones). Close the tap and pour in the organic solution (using a funnel). Add the water (or aqueous acid or base) and insert the stopper.

SAFETY It is sensible to wear rubber gloves if acids or bases are used.

Holding the stopper in place with the heel of one hand and supporting the closed tap with the other, invert the funnel and gently swirl the contents around (do NOT shake) for a few seconds. With the funnel still inverted, open the tap to release the pressure (Fig. 3.3b) (the pressure build-up may be quite high with volatile solvents such as ether). Close the tap, swirl more vigorously and release the pressure again. Do this repeatedly until there is no pressure build-up. After that the separating funnel can be shaken to mix the contents, but do not shake too vigorously or too long or you will produce an emulsion that will be very difficult to break.

SAFETY Don't shake the funnel close to your face.

Then support the funnel with a clamp or ring and allow the two phases to separate until a sharp interface is produced. This may take a long time if the shaking has been overvigorous. Sometimes coagulation of the droplets can be encouraged by gentle swirling. If an emulsion has formed, it can sometimes be broken by adding a few drops of an alcohol such as pentanol or octanol or a small amount of neutral electrolyte such as sodium chloride (which will dissolve in the aqueous layer). However, patience is needed here or the separation of the two layers will be difficult and product will be lost. When the layers have separated the stopper is removed and the lower layer is run off into another vessel. The lower layer may be organic or aqueous depending on the density of the organic solvent used (check if you are unsure by taking out a few drops with a dropper and finding out if it is water-miscible). The extraction procedure is usually repeated several times and the extracts are combined for further processing. It is important to keep both layers in all cases until the work-up is complete.

SALTING OUT. In some cases aqueous extractions are complicated by the fact that the desired organic compound is slightly water-soluble and therefore some of it finishes up in both layers when ideally it should be wholly in the organic one. In such cases the aqueous solubility of the organic compound can be markedly reduced by the presence of dissolved salts, usually sodium chloride for neutral or basic materials or ammonium sulphate for acidic ones. This effect can be used in several ways in extractive work-up procedures and the experimental instructions will usually specify if salting out is required.

(b) Continuous extraction

Extraction with a separating funnel is satisfactory only if the partition coefficient of the compound concerned is so high that it partitions almost completely into one layer or the other. However, some organic compounds (fortunately relatively few), such as carboxylic or sulphonic acids, are readily soluble in both organic solvents and

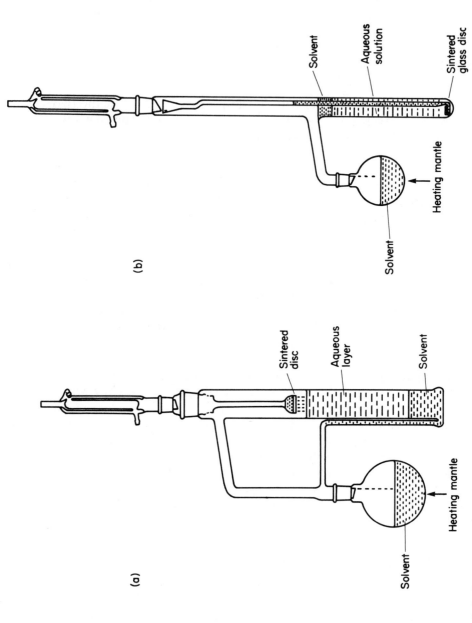

Fig. 3.4 Continuous liquid/liquid extractors: (a) for solvents heavier than water; (b) for solvents lighter than water.

water. It is only possible to extract such materials from aqueous solution by a continuous extraction procedure, for example using a piece of apparatus such as that shown in Fig. 3.4a. This is suitable for solvents that are heavier than water (e.g. dichloromethane). The solvent is continuously distilled into the condenser, from where it falls through the aqueous solution as fine droplets to collect at the bottom for return to the flask. As it passes through the aqueous layer the solvent extracts a little of the desired compound, which then accumulates in the flask as the extraction proceeds. An equivalent system is available for solvents lighter than water (Fig. 3.4b).

(c) The extraction of solids

It is sometimes necessary to extract a solid organic product from a solid mixture containing other materials (usually inorganic) that are soluble in neither organic solvents nor water. This is easier to do if the mixture is first ground up finely with a mortar and pestle. If the organic solid is readily soluble in an organic solvent then the separation is easily achieved, e.g. by putting the mixture into a sintered funnel (Fig. 3.11), adding the solvent, stirring and then applying suction. However, if the organic solid is not easily soluble then a continuous extraction method, using a Soxhlet extractor (Fig. 3.5), must be used. The mixture is placed in the porous thimble. In operation, the solvent distils into the extraction chamber and, when full, siphons back into the flask. Prolonged repetition of this cycle slowly transfers the solid into the flask.

3.1.4 Drying organic solutions

After an aqueous extraction procedure the organic layer will contain some water both as small droplets and as dissolved water (many solvents can dissolve appreciable amounts of water; see Table 3.2, p. 82). This may be removed by adding a little powdered inorganic desiccant to the solution to absorb the water and then filtering to remove the desiccant. Many desiccants have been used for this purpose but it has recently been shown[1] that anhydrous calcium chloride, anhydrous magnesium sulphate and powdered molecular sieve are particularly effective for the rapid (15–30 min) drying of grossly wet ether solutions. Calcium chloride, however, is of limited use as it reacts with many types of compound and so cannot be used for organic acids, alcohols, phenols, amines, amides, aldehydes, ketones and

1. Burfield, D. R. and Smithers, R. H. (1982) *J. Chem. Educ.*, **59**, 703.

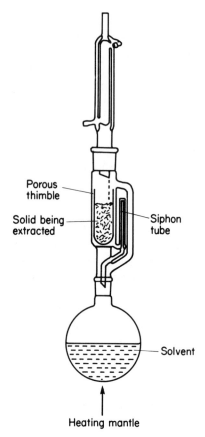

Porous thimble

Solid being extracted

Siphon tube

Solvent

Heating mantle

Fig. 3.5 Soxhlet continuous extractor for the extraction of solids.

esters. Powdered molecular sieve (type 4A) is very efficient (see below), and can be used for almost all types of compound, but is expensive. This leaves anhydrous magnesium sulphate as the first choice as a general-purpose drying agent; it is cheap, effective and unreactive towards most functional groups.

The procedure is to place the organic solution in a conical flask and add *c*. 1 g of magnesium sulphate (or the other desiccants) per 10 ml of solution. The mixture should be left for 15–30 min with occasional stirring or shaking and then filtered by suction (Section 3.2.3) through a dry sintered funnel. Since drying is usually followed by rotary evaporation (Section 3.1.2) it is often convenient to filter directly into a suitable size RB flask (Fig. 3.11c).

If it is important to get the water content down to a really low level then it is best to carry out primary drying as above and then to dry the solution further using anhydrous calcium sulphate ('Drierite') or molecular sieve (type 4A or 5A). Molecular sieve acts rapidly in powdered form (e.g. reducing the water content of water-saturated ether to 0.092 mg per gram in 15 min) while the granulated form is much slower (e.g. taking 6 h to reduce the water content of ether to 0.29 mg per gram).

3.1.5 Separation of the target product(s)

The primary work-up procedure will have removed any inorganic material and organic solvents to leave the crude organic product(s). This material should be weighed.

(a) Reactions giving a single major product

If the reaction is a straightforward one, it will most likely be possible to obtain the target compound by crystallization or by distillation as specified in the instuctions for the experiment. However, problems sometimes arise with products that should solidify when the solvent is finally removed, but do not. In such cases what is often obtained is a viscous oil that refuses to solidify even though the solvent has been removed as thoroughly as possible on a rotary evaporator. Leaving the oil to stand at room temperature overnight will often produce crystals, but if not there are several stratagems that can be tried to induce crystallization. It is usually better to try them on a small sample of the product (in a small test tube) and, if successful, to use the solid obtained to 'seed' the bulk material. The most common method is by scratching the inside of the vessel at the surface of the oil with a glass rod. This is intended to provide nucleation sites, which may initiate crystal growth. If scratching at room temperature is not effective then it sometimes helps to cool the sample to −80°C or even in liquid nitrogen and allow it to warm up slowly. Trituration or 'rubbing' the oil with light petroleum or hexane is particularly effective for syrupy mixtures. The oil is covered with a little of the solvent and rubbed with a glass rod against the bottom of the vessel. In favourable cases the solvent dissolves out traces of impurities that may be preventing crystallization while the grinding action provides nucleation sites.

If these methods fail then it may be necessary to take a small amount of the mixture and separate it by chromatography (see next chapter) to produce a small sample of the pure compound, which should crystallize without difficulty. This can then be used to 'seed' the bulk sample. This method is often quicker and cheaper than carrying out chromatography on the whole product.

If none of these methods work it must be concluded that the 'solid' is too impure to crystallize and it must be separated out by chromatography.

(b) Reactions giving mixtures

Mixtures of liquids can usually be separated by fractional distillation (Section 3.4.3) but in some cases distillation is not practicable because either the amount of material is too small or the liquid is too involatile or too sensitive to heat. Small amounts (<1 ml) of 'volatile' liquids can be separated by preparative-scale gas–liquid chromatography (Section 4.1.2). This method can effect separations that are impossible by distillation and is invaluable for obtaining small samples of pure material for spectroscopic examination. However, it is essentially a small-scale technique and not generally suitable for 'preparative-scale' separations. Involatile liquids that cannot be distilled can be separated by the chromatographic techniques used for solids.

Mixtures of solids are usually separated by some form of liquid/solid (adsorption) chromatography. Such mixtures should first be examined by thin-layer chromatography (TLC) (Section 4.1.1) to find out how many components are present and how easy they are to separate. The preparative-scale separation can then be carried out by 'flash', 'dry-column flash', or 'medium-pressure' chromatography as appropriate (Section 4.2.2).

3.2 CRYSTALLIZATION

Crystallization is the most common method for the purification of organic solids that are not heavily contaminated with other substances*. It is one of the most frequently performed operations in practical organic chemistry and, although it is not basically a difficult technique, it does need much practice to do it well – particularly when working on a small scale.

3.2.1 General principles

The technique makes use of the knowledge that solid compounds are much more soluble in hot solvents than in cold ones. Thus if you pre-

---

* In cases where appreciable quantities of other compounds (c. 5–10% or more) are present, it is usually necessary to separate the mixture by chromatography (Chapter 4).

pare a saturated hot solution of compound A and allow it to c
solution will become supersaturated and the compound will se
out as crystals. If the compound is impure, say it contains a fe
cent of another compound B, then the impurity will also disso ... in
the hot solvent, but when it cools down the solution will not be super-
saturated with compound B (because it is present in low concentration) and it will stay in solution while the major component, compound
A, crystallizes out. Thus the pure, crystalline compound A can be
filtered off while the impurity B stays in solution in the filtrate (generally known as the mother liquor).

In practice the method involves the following five steps: (i) dissolving the solid in the minimum volume of a boiling solvent; (ii)
filtering the hot solution to remove insoluble impurities (if any are
present); (iii) allowing the solution to cool so that the solid will crystallize out, and then to stand until crystallization is complete; (iv)
separating the crystals from the mother liquor by filtration; and (v)
drying the crystals. In step (iii) the minor components (soluble impurities) stay in solution while the major component crystallizes out. It
may be necessary to recrystallize the material several times to achieve
absolute purity.

The key to success in crystallization lies in using the best solvent,
i.e. one that will dissolve the material easily when hot, but in which   ✗ importa
the major component is almost insoluble when cold, so allowing
most of it to crystallize out. In teaching exercises in the early stages
the crystallization solvent will be specified, but in more advanced
work the procedure for the selection of a suitable solvent (Section
3.2.4, p. 80) is an essential preliminary which requires careful work
and good judgement.

3.2.2 Melting point as a criterion of purity

Melting point (m.p.) provides a useful indication of the purity of a
solid (Section 3.3) since the presence of impurities lowers the m.p.
and widens the melting range. It should be measured for the material
both before and after crystallization. The compound may be judged
to be pure when the m.p. reaches a maximum value, i.e. it is un-
changed by further recrystallization. This is known as 'crystallization
to constant m.p.'

SAFETY Most solvents used in crystallization are volatile and
flammable and some are toxic. Take appropriate safety precautions
(Section 1.3) to prevent inhalation or ignition.

3.2.3 Methods of crystallization

The procedure to be used depends on the amount of sample. The method in subsection (a) below can be used for any amount down to *c*. 100 mg or even less. However, at the lower end of the range, say 500 mg or less, it is easier to use the method in subsection (b) p. 78, particularly if hot filtration is required. This method can be used for amounts down to *c*. 10 mg.

(a) Crystallization using flask/funnel techniques

If the solvent to be used is not specified then go through the procedure in Section 3.2.4 (p. 80) first.

(i) Dissolving the solid in the boiling solvent

The objective here is to dissolve the solid in the minimum volume of boiling solvent in a conical flask*. It is essential to use a flask with a ground-glass socket so that a reflux condenser can be fitted (Fig. 3.6). These flasks are available in sizes down to 5 ml. The size chosen should be such that it will not be more than *c*. half-filled with solid and solvent during the operation. The size required may be judged from the preliminary solvent selection exercise or guessed from experience, e.g. *c*. 1 g of solid would require a 25 or 50 ml flask.

PROCEDURE. Weigh out the amount of material to be crystallized and place it in the flask, using a powder funnel (Fig. 2.5). Add a little of the solvent – just enough to cover the solid but not enough to dissolve all the solid at boiling point. Add a few anti-bumping granules. Fit a reflux condenser (Section 2.1.4), using an adapter if necessary.

SAFETY The use of a reflux condenser is essential, whatever the scale of operation. It is not safe to heat any solvent in an open vessel.

Set up the flask and condenser on a water bath for solvents of b.p. up to *c*. 80°C (Fig. 3.6) or an electric hotplate for higher-boiling solvents. Heat the mixture until the solvent is boiling gently under reflux. After a few minutes, some but not all of the solid should have dissolved. Using a dropping pipette add more solvent, a little at a time, down the condenser until the solid has completely dissolved. It is important to do this slowly, allowing the solvent to boil for a few

---

* Crystallization should never be carried out in a beaker or any other open vessel. A conical flask allows easy removal of the solid after crystallization whereas a round-bottomed flask does not.

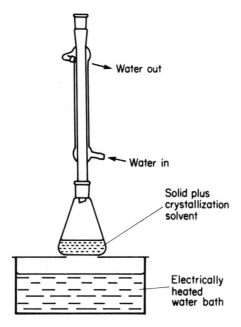

Fig. 3.6 Apparatus for crystallization.

minutes after each addition to give the solid time to dissolve. Remember that the intention is to use the minimum volume of solvent. Record the volume of solvent used.

If the solution is clear (no suspended solids) and is not highly coloured by tarry impurities then it should be set aside* to crystallize (subsection (iii) below).

However, there are two situations where further action is needed. The first is where the solution contains insoluble material, e.g. dust or traces of inorganic material. This must be removed by filtering the hot solution (see next subsection). The second is where the sample is contaminated by highly coloured, tarry impurities (as may be the case if the solid is the crude product from a reaction). These contaminants may be removed by allowing the solution to cool a little, adding powdered charcoal (*c.* 1–2% of the weight of the organic solid), and then heating the solution to reflux again for a few minutes.

charcoal

SAFETY Do not add charcoal directly to the boiling solution as it may cause violent bumping.

---

* The anti-bumping granules can be removed with a spatula or by decanting the solution into another conical flask.

The charcoal will absorb the impurities (in most cases) and is then re-moved by hot filtration (next subsection).

(ii) Filtering the hot solution

This step is required ONLY if insoluble impurities are present. It is the step that causes most problems for inexperienced workers (and sometimes for experienced ones as well). The difficulty arises if the solution is allowed to cool down during filtration; the solid may then start to crystallize in the filter funnel, where it will block any further filtration. This problem is less likely to occur with a solution that is not absolutely saturated when hot, so, if hot filtration is required, an extra *c.* 5% of solvent should be added at the conclusion of sub-section (i) above.

Several procedures are available. The first one described below, suction filtration using a pre-heated funnel, works well for operations on a small to medium scale (up to say 50–100 ml). The second method, gravity filtration, is generally used for larger-scale operations and usually employs a device for keeping the filter funnel hot.

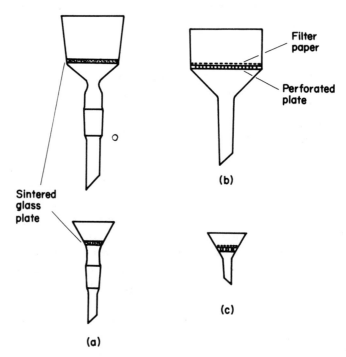

Fig. 3.7 (a) Filter funnels with a sintered glass plate; (b) Büchner funnel; (c) Hirsch funnel.

SUCTION FILTRATION USING A PRE-HEATED FUNNEL. This is a quick and effective technique that works well for solvents other than the very volatile ones (ether, 40/60 petrol and dichloromethane – Section 3.2.4). The filtration can be carried out using either a sintered glass funnel (Fig. 3.7a) or a Büchner- or Hirsch-type funnel fitted with a filter paper (Fig. 3.7b and c). The latter type is essential for charcoal filtration (see last part of this section). The key requirement for sintered funnels is that the sinter is clean and free-flowing so that the filtration can be done quickly with the minimum suction. Partially blocked sinters cause slow filtration, cooling of the solution and failure of the operation. It is therefore important to check the sinter before use by 'filtering' some pure solvent under similar conditions. To avoid this problem many people prefer to use a Büchner or Hirsch funnel with filter paper.

A funnel of appropriate size should be used. The type with an integral side arm (Fig. 3.10) is particularly convenient for small volumes. Filtration can be either directly into a conical flask (Fig. 3.8a) or, for small volumes, into a small test tube (Fig. 3.8b).

The techniques with a sintered funnel and with filter paper differ slightly as described below.

Procedure using a sintered funnel. Pre-heat the funnel in an oven so that it is at a temperature just above the b.p. of the solvent. Transfer it rapidly to the filter flask or tube (Fig. 3.8), pour in the hot solu-

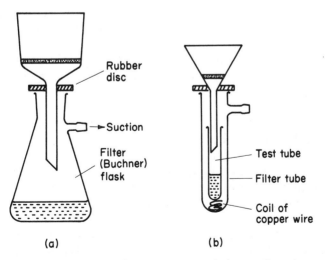

Fig. 3.8 Filtration apparatus for (a) large- and (b) small-scale work.

tion and apply a very light suction to the flask (use a water pump with the air inlet tap partially open, p. 111).

The suction should be just enough to keep the solvent flowing quickly through the funnel but not enough to cause rapid evaporation and cooling of the solvent. Keep adding more hot solution to keep the sinter covered with liquid until the end of the operation. The filtrate should be warmed to redissolve any crystals that may have appeared, and then allowed to cool slowly.

Procedure using a Büchner or Hirsch funnel with filter paper. This requires a slight variation of the above technique. Heat the funnel as described above, transfer it to the filter flask (Fig. 3.8), fit a filter paper and moisten it with hot pure solvent (heated separately) so that it sticks to the funnel. Then apply a light suction (use a water pump with the air inlet tap partially open) and immediately pour the hot solution into the centre of the funnel. It is important that the suction is applied before the liquid is poured in, to prevent the filter paper floating off the perforated base. Then proceed as in the second paragraph of the previous subsection above.

Filtration of solutions containing charcoal or other colloidal impurities. Charcoal contains very fine particles that can irreversibly clog glass sinters. Always use the filter paper method described above. However, filter paper, used by itself, can also be rapidly choked up by the charcoal. This can be prevented by using a filter aid such as Celite. This is a fine powder that should be added to the solution (in an amount *c.* 0.5% of the solute) just before filtration, or placed in a layer (*c.* 3 mm thick) on top of the moistened filter paper and then itself moistened with hot solvent.

The same technique can be used to allow rapid filtration of either hot or cold solutions contaminated with other very fine or colloidal impurities.

GRAVITY FILTRATION. This is generally used for larger volumes of solution. It requires the use of a short-stemmed glass funnel (Fig. 3.9) fitted with a fluted filter paper (ask your instructor to show you how to fold a fluted filter paper). This method of folding gives much faster filtration than the normal method. It is of course essential to keep the liquid level below the top of the filter paper. Relatively large filter papers are required compared to those used in Büchner and Hirsch funnels and these soak up quite a lot of solution, making the method unsuitable for small volumes. On a small scale the funnel can simply be pre-heated in an oven to a temperature just above the boil-

Fluted filter
paper

Funnel
heater

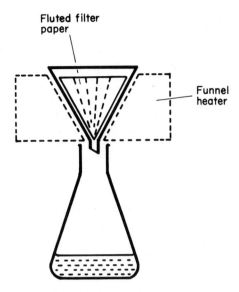

Fig. 3.9 Gravity filtration of a hot solution using a fluted filter paper.

ing point of the solvent, but if a large volume is to be filtered, some device is required to keep the funnel and its contents hot (Fig. 3.9). Various kinds of heater are available; some are filled with hot water and others are electrically heated.

SAFETY This operation must be carried out in a fume-cupboard and there should be no sources of ignition, flames or electrical devices in the vicinity while filtration is being carried out.

(iii) Formation of the crystals

The saturated (or nearly saturated) hot solution is allowed to cool down to room temperature and, as it does so, the crystals are usually deposited. This can be a slow process and the solution should be allowed to stand at room temperature for some time and observed carefully until crystallization is complete. In general slow cooling will produce large crystals and rapid cooling will give very small powdery ones, but this varies enormously depending on the nature of the compound and the solvent. Rapid cooling can lead to the formation of oils (see below).

When crystallization is complete, the crystals are filtered off, dried and weighed (next section). If the recovery of purified material is low it indicates that the solvent used was not ideal (or too much was used). In such cases a further crop of crystals may be obtained by cooling the filtrate (mother liquor) below room temperature in an ice/water bath, or to an even lower temperature in a deep freeze or cooling bath (Tabe 2.1). In cases where the compound is of particular value the mother liquor can be concentrated using a rotary evaporator (Section 3.1.2), and cooled to produce yet another crop. In general these later crops will be less pure than the main crop and should be kept separate until their purity has been determined by melting point or some other method.

PROBLEMS: NO CRYSTALS FORM. In some cases crystals will fail to appear as the solution cools even though the solution is super-saturated. This is usually because of the lack of suitable nuclei to initiate crystal formation and is most often found with the dust-free solutions obtained after hot filtration. Several techniques can be used to induce crystallization. The best method is to add a 'seed' crystal of the same substance (it is useful to keep a small sample of the crude material for this purpose). An alternative is to scratch the inside of the flask, at the level of the liquid, with a sharp glass rod. This pro-duces irregularities in the glass surface which act as sites for crystal growth. Further cooling of the solution sometimes helps, but with some solvents extreme cooling produces viscous solutions that will never crystallize. If these immediate methods fail then it is sometimes effective simply to stopper the flask, label it and leave it in your bench cupboard or the refrigerator for a few days.

PROBLEMS: FORMATION OF OILS. In the case of low-melting substances, the material may separate as an oil instead of crystals, particularly if the solution is cooled rapidly. Once the oil has started to form, impurities or some of the solvent may dissolve in it and de-press the melting point, so the oil may persist at temperatures well below the expected m.p. of the compound. Even if it does solidify on standing, it is likely to give very impure material.

In such cases it is necessary to experiment with dilution, seeding and very slow cooling to prevent oil formation in the first place. Thus in a case where an oil starts to separate as the solution cools you should reheat the solution until the oil just disappears and then allow it to cool very slowly while stirring gently with a glass rod. If it forms an oil again, then reheat as before, allow to cool a little and then add a few 'seed' crystals or scratch the flask with a glass rod (see above) as it cools further. If the seed crystals dissolve before the oil or solid has appeared add a few more. If this is not effective then reheat again and add a little more solvent and repeat the procedure. Much patience and persistence is often called for to obtain crystals and even then the recovered yield may not be very high. The crystallization of low-melting substances is made much more difficult if they contain appreciable amounts of impurity, and in such cases it is often better to purify the substance by chromatography (Chapter 4) before crystallization is attempted.

(iv) Filtering off the crystals

When crystallization is complete the crystals are separated from the cold mother liquor by suction filtration. Two types of filter funnel are in general use: (a) those with a sintered glass plate (e.g. Fig. 3.7a), which need no filter paper (the sinters are available in various degrees of porosity from 0 (coarse) to 5 (fine), porosity 3 being the most common for general use); and (b) the Büchner- and Hirsch-type funnels (Fig. 3.7b and c), which have a perforated plate that supports a filter paper. Both types are available in a wide variety of sizes. The ones with a sintered glass plate, generally known simply as 'sintered funnels', are better as the solid can be gently scraped out after filtration with no risk of breaking up the filter paper and contaminating the crystals. Very small amounts of solid (20 mg or less) can be filtered off using the smallest sintered funnels (Fig. 3.10), obtainable with discs of diameter down to c. 7 mm (alternative methods for small amounts are discussed on p. 78). It is important that these sinters are kept clean and free-flowing (consult your instructor or laboratory technician about cleaning filters that are heavily clogged).

The filter funnel is attached to a filter (Büchner) flask (Fig. 3.11a) or a filter tube (Fig. 3.11b) either via a cone/socket joint (Fig. 3.11) or

Fig. 3.10 Sintered funnel for very small-scale work.

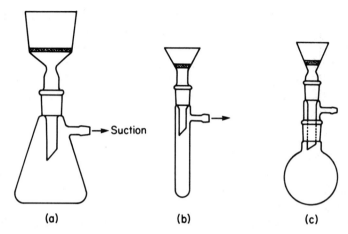

→ Suction

(a)   (b)   (c)

Fig. 3.11 Equipment for filtering off crystals.

using a flexible rubber disc (Fig. 3.8), which provides an adequate vacuum seal when the funnel is pressed gently down. Alternatively a side-arm adapter (Fig. 3.11c) can be used, with either a conical or a round-bottomed flask. The latter is particularly useful if the filtrate is to be concentrated later on a rotary evaporator.

Suction is provided by a water-pump-driven vacuum system (Section 3.4.8).

PROCEDURE*. If a Büchner or Hirsch funnel is being used it should be fitted with a filter paper of size sufficient to cover the perforated plate but not to fold up against the sides. The filter paper should be moistened with a little of the pure solvent so that it sticks to the perforated plate. Apply gentle suction to the system. Swirl the crystals/mother liquor round gently in the crystallization flask and pour the mixture smoothly into the funnel. Use only sufficient suction to maintain a smooth flow through the filter. When filtration is complete, release the vaccum and if necessary pour part of the mother liquor back into the crystallization flask to rinse out any crystals adhering to the walls. When all the crystals have been transferred they may be washed with a little of the cold pure solvent. To do this, release the suction, add just enough solvent to cover the crystals, stir gently and briefly with a spatula and apply the suction again. Washing the crystals in this way should be done with much circumspection

---

* See p. 75 for the filtration of solutions that have been cooled below 0°C.

as it may entail considerable loss if the crystals are appreciably soluble in the cold solvent.

Finally, cover the filter funnel with a watch (or clock) glass to keep out dust and apply suction to remove as much solvent as possible. For moderate and large amounts of crystals it is useful to press them down on the filter with a wide glass stopper during this operation, but this is not appropriate when working on a small scale as it may entail loss of material.

FILTERING CHILLED SOLUTIONS. In some crystallizations it is necessary to cool the mixture below ambient temperature to obtain a good recovery of material. In such cases it is an advantage to cool the filter funnel before use to avoid warming the solution during filtration. The cooling is no problem; it can be done easily by putting the funnel, preferably in a polythene bag to keep it clean and dry, in a refrigerator or deep freeze. The difficulty is condensation – on the funnel, the solution and the crystals. It is almost impossible to avoid it completely, but, unless the compound is water-sensitive, it does not really matter as the crystals can be dried (see below) after filtration. The best that can be achieved without elaborate arrangements is to set up the funnel rapidly under a stream of dry nitrogen (Fig. 3.12) and keep it covered with a clock glass during filtration.

Craig tubes (p. 80) offer considerable advantages for low-temperature crystallization of small amounts since the transfer operation is eliminated.

(v) Drying the crystals

AT ROOM TEMPERATURE. Most samples can be dried effectively at room temperature by using a vacuum to remove the residual traces of crystallization solvent. After removing as much of the solvent as possible by suction in the filter funnel, transfer the crystals to a weighed crystal dish (Fig. 3.13) and cover with a watch or clock glass. Select a dish of such a size that the crystals are spread out in a fairly thin layer. Very small samples can be dried in a sample tube capped with a piece of aluminium foil perforated with a few pinholes. Place the dish (or tube) in a vacuum desiccator (Fig. 2.13). Sample tubes should be supported in a small beaker. Put a guard over the desiccator and evacuate it. Most solvents of b.p. *c.* 80°C or less are readily removed under water-pump vacuum (Section 3.4.8(a)) but less volatile ones such as toluene, propanol, butanol, etc., require the higher vacuum obtainable with a rotary oil pump (Section 3.4.8(b)). After drying for 0.5–1 h, release the vacuum (see p. 26), weigh the sample and determine its m.p. Since small residual traces of solvent

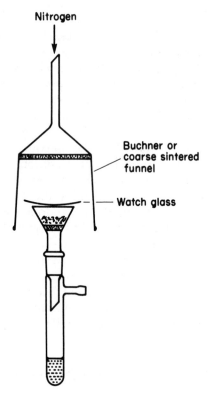

Fig. 3.12 Filtration under nitrogen.

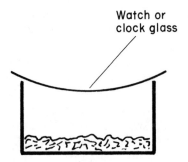

Fig. 3.13 Drying crystals in a crystal dish.

can depress the m.p., it is usually worth while to repeat the drying procedure for a further 0.5 h and determine the m.p. again.

AT ELEVATED TEMPERATURE. Occasionally it may be necessary to dry the crystals at an elevated temperature, e.g. when a very involatile crystallization solvent has been used or when preparing an analytical sample (Section 3.2.5(a)). This is again done under vacuum* and obviously at a temperature well below the m.p. (or sublimation temperature) of the sample. Large amounts of material can be dried in a crystal dish placed in a vacuum oven. Smaller samples are dried in a boat in a drying pistol (Fig. 3.14). Very small samples can be dried in a sample tube capped loosely with aluminium foil perforated with a few pinholes and placed in the boat.

The chamber of the drying pistol can be heated either by the vapour of a refluxing solvent, Fig. 3.14 (see Table 3.2 for boiling points) or electrically. In all cases the sample is placed in the oven or pistol when cold, evacuated to the required vacuum (usually oil-pump vacuum (Section 3.4.8(b)) and then the heating is applied.

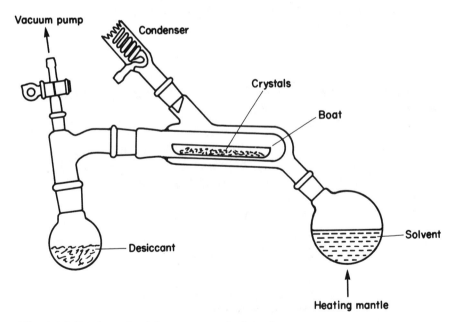

Fig. 3.14 Drying pistol for vacuum drying above room temperature.

* NEVER in a heated laboratory oven at atmospheric pressure, for many reasons – melting, thermal decomposition, toxicity hazard, contamination.

Removal follows the reverse procedure using the same technique for admitting air as for a vacuum desiccator (Section 2.1.5).

(b) Crystallization of small samples

Much preparative work in organic chemistry is carried out on a small scale, giving products in amounts of 500 mg (0.5 g) or less. The crystallization of amounts down to 100 mg can be carried out using the conventional flask/funnel methods described in the last section, using very small apparatus. Even smaller amounts can be crystallized in tiny conical flasks or micro test tubes if hot filtration is not required. However, when working on a very small scale*, e.g. when preparing samples for elemental analysis or spectroscopy, absolute purity is essential and hot filtration to remove dust is strongly recommended. Amounts of 100 mg or less are most conveniently handled using the combined flask/filter apparatus shown in Fig. 3.15. This apparatus minimizes handling and transfer losses, which can be considerable when using conventional methods for small-scale work. It can be conveniently used for amounts up to c. 500 mg. Useful sizes are 10, 5 and 2 ml. A rubber-bulb air pump equipped with a cone to fit the side arm is also required.

(i) Dissolving the sample and hot filtration

SAFETY This operation should be carried out in a fume-cupboard away from any source of ignition.

Introduce the sample into the bulb through the side arm using a micro-spatula†. Add one or two anti-bumping granules and then a little of the crystallizing solvent, with a dropping pipette, using it to wash down any sample adhering to the side arm. Clamp the device where shown and heat the bulb gently over a small water bath (or better, in a small oil bath heated by a hotplate (Fig. 2.7). If a water bath is used the 'spout' can be protected from condensation with a piece of aluminium foil crimped on lightly or with a clean dropper teat. Adjust the

---

* Milligram quantities are often more easily purified by sublimation (Section 3.4.7), but this does not work for all compounds.
† The handling of small quantities of fine powder or crystals is often made impossible by the build-up of static electricity, which causes the solid to adhere to glass surfaces and sometimes to fly off the spatula in all directions during transfer. This problem can be overcome by using an anti-static pistol (e.g. Aldrich 'Zerostat') immediately before transfer.

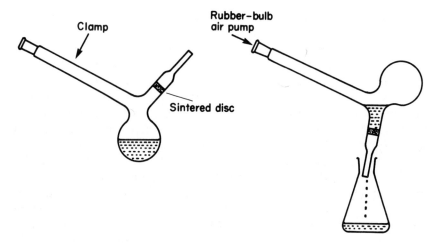

Fig. 3.15 Apparatus for small-scale crystallization with hot filtration.

heat so that the solvent refluxes gently up to the level of the filter disc (but not above). The long side arm will serve as a reflux condenser. Add more solvent gradually down the side arm until the compound has dissolved. Continue to reflux gently until the filter disc has been warmed up by the solvent vapour. Remove from the heat, quickly wipe off any oil or water from the outside of the bulb, place a small conical flask or a Craig tube (Fig. 3.16 and see below) over the spout, rotate the apparatus (Fig. 3.15) and apply light pressure with a rubber bulb to drive the hot liquid through the filter into the receiver. Hot filtration of small volumes under pressure is far easier than under vacuum since the evaporation/cooling problem is avoided.

The sudden cooling of the solution when it reaches the receiver may induce rapid crystallization. If so, the solution should be re-heated until the crystals have dissolved and then allowed to cool slowly.

(ii) Isolating and drying the crystals

For amounts in the 500–50 mg range the hot solution may be filtered into a small conical flask or a micro test tube and allowed to crys-tallize in the usual way (Section 3.2.3). The crystals are then filtered off in a small sintered funnel of the type shown in Fig. 3.10. How-ever, for amounts of 100 mg or less the use of a Craig tube (Fig. 3.16), as described below, much reduces the handling and transfer losses.

USE OF THE CRAIG TUBE. Filter the hot solution into a Craig tube of appropriate size, as described above, fit the loose stopper and allow to stand until crystallization is complete (Fig. 3.16a). Then place a centrifuge tube over the Craig tube and invert the assembly (Fig. 3.16b). Centrifuge for 1–2 min to drive the mother liquor off the crystals. Carefully pour off the mother liquor and return the Craig tube to an upright position. Tap the tube to detach any crystals from the stopper. If desired the crystals can be washed by loosening the stopper, rinsing the crystals down with a few drops of chilled solvent, and centrifuging again. The crystals are dried by removing the stopper, loosely capping the tube with aluminium foil and placing it in a drying pistol (Fig. 3.14). If further recrystallization is required, it is carried out in the Craig tube by adding filtered solvent and gently heating the base of the tube in a water or oil bath (in a fume-cupboard) to redissolve the crystals; the tube itself will act as a reflux condenser. The cycle can be repeated as often as required and incurs no transfer losses.

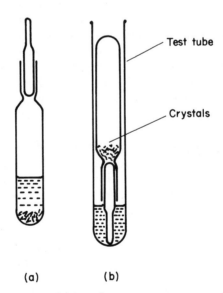

Test tube

Crystals

(a)   (b)

Fig. 3.16 'Filtration' in a Craig tube.

3.2.4 Choosing the crystallization solvent

The success of crystallization depends largely on the proper choice of solvent. The ideal solvent is one in which the substance to be purified is readily soluble when hot, but only sparingly soluble in the cold, and in which the impurities are very soluble.

(a) General considerations

In practice, the choice of solvent is made by a series of trial crystallizations (following section (b)), but some guidance may be obtained from the generalizations below. The most useful guide is that 'like dissolves like', i.e. polar compounds are more soluble in polar solvents than in non-polar solvents and vice versa. Some general solubility characteristics and an indication of solvent polarity are given in Table 3.1 and some values of solvent dielectric constants (which indicate polarity) are given in Table 3.2. For example, a polar compound containing hydroxy (OH) groups is likely to be soluble in methanol (CH_3OH) but to be progressively less soluble in ethanol (C_2H_5OH) and the higher alcohols, which get less polar as the length of the hydrocarbon chain increases. Highly polar compounds with, for example, several hydroxy (OH) groups, or a carboxyl group (CO_2H), or a sulphonic acid group (SO_3H), tend to be at least partially water-soluble but this obviously depends on what else is present in the molecule. Water solubility is opposed by the presence of hydrophobic groups like hydrocarbon chains or rings. In general most organic molecules, apart from small ones containing OH, CO_2H, or SO_3H groups, are not water-soluble and it is rare that water is a good crystallization solvent. Non-polar compounds like hydrocarbons or alkyl halides

Table 3.1 General solubility characteristics of some classes of compound

Type of substance	Polarity	Good solvents
hydrocarbons	low	pentane, hexane, light petroleum, toluene
ethers		diethyl ether
alkyl halides		chloroform
esters		
aldehydes and ketones		acetone
phenols		ethyl acetate, dichloromethane
alcohols		ethanol
carboxylic acids		
sulphonic acids	high	water

Table 3.2 Commonly used solvents for crystallization

SAFETY Most of these solvents are flammable and some are toxic by inhalation. Observe proper safety precautions (Section 1.3, p. 3).

Solvent[a]	B.p. (°C)	Dielectric constant	Water solubility[b] (g/100 g)
pentane	36	2.0	0.03
hexane	69	1.9	insol.
light petroleum	60–80	c. 2	insol.
cyclohexane	81	2.0	sl.sol.
toluene	110	2.4	sl.sol.
diethyl ether	35	4.3	7.5
ethyl acetate	77	6.0	9.0
acetic acid	118	6.2	misc.
dichloromethane	40	9.1	2.0
propan-2-ol	82	18	misc.
acetone	56	21	misc.
ethanol	78	25	misc.
methanol	65	34	misc.
dimethylformamide	154	38	misc.
dimethylsulphoxide	189	45	misc.
water	100	80	–

[a] Those solvents in **bold** type are recommended as first choices for trial crystallization (see text).
[b] Insoluble (insol.), slightly soluble (sl.sol.) or miscible (misc.).

are virtually insoluble in water but dissolve readily in non-polar solvents such as light petroleum* or toluene.

CHOICE OF SOLVENT BOILING POINT. In general, it is best to pick a solvent with a wide temperature range so that the solubility difference between cold and hot will be as large as possible. For this

---

* Light petroleum, sometimes called 'petroleum ether', is a mixture of hydrocarbons, NOT an ether, and is available in several boiling-point ranges, e.g. 40–60°C, 60–80°C, 80–100°C. These fractions are usually referred to as '40/60 petrol', '60/80 petrol', etc. All are non-polar and highly flammable.

reason low-boiling solvents such as ether (diethyl ether), 40/60 petrol and dichloromethane should be avoided where possible. The use of high-boiling solvents also has disadvantages in that they are much less volatile and are difficult to remove from the crystals after filtration. A solvent whose b.p. is higher than the m.p. of the substance to be crystallized should be avoided since the compound will melt before it dissolves and on cooling will probably come out as an oil rather than crystals.

The most popular solvents* are those with b.p.'s in the 60–90°C range such as: 60/80 petrol, hexane, cyclohexane, ethyl acetate and the lower alcohols (up to C_3 or C_4), as set in bold type in Table 3.2. If the compound is not soluble in any of these, then higher-boiling solvents such as acetic acid (glacial), pyridine, the higher alcohols, dimethylformamide or dimethylsulphoxide should be tried.

(b) Trial crystallizations

The solvents given in bold type in Table 3.2 are recommended as first choices for trial crystallization. A little of the solid (c. 50 mg)[†] is placed in a clean test tube (7.5 × 10 mm) and 0.25–0.5 ml of solvent is added dropwise with shaking. If the solid dissolves in the cold the solvent is obviously unsuitable, except as the 'good' solvent of a mixed solvent pair (see below). If it does not dissolve then heat the tube, with shaking, over a water bath until the solvent boils. If all the solid has not dissolved, add more solvent, a little at a time until c. 1.5 ml has been added. If some of the solid still remains undissolved then the low hot-solubility makes the solvent unsuitable and another should be tried. If a clear solution is obtained, allow the solution to cool. If no crystals are obtained after the solution has been at room temperature for a few minutes, add a 'seed' crystal or scratch with a glass rod (Section 3.2.3). If the solvent is a good one, then a good crop of crystals will form. If the recovery is poor then the solvent is unsuitable because of high cold-solubility. Repeat the process with several solvents and select the one that produces the largest crop of crystals[‡].

MIXED SOLVENTS. In some cases it will be found that no single solvent is suitable. It is then necessary to use a mixture of two solvents,

---

* Previously popular solvents (carbon tetrachloride, benzene and chloroform) are highly toxic or carcinogenic and should not be used for crystallization.
† For small-scale work this can be reduced to c. 10 mg and the amount of solvent reduced in proportion.
‡ If the compound is virtually insoluble in all suitable solvents see Section 3.2.5.

one in which the substance is readily soluble and the other in which it is only very sparingly soluble. The two solvents, of course, must be completely miscible, and if possible should have similar boiling points. Two methods of procedure can be followed: (i) suspend the solid in a little of the 'poor' solvent, heat to boiling and then add the 'good' solvent dropwise until the solid dissolves; or (ii) dissolve the solid in a little of the 'good' solvent at reflux and add the 'poor' solvent dropwise until a faint opalescence appears and then add a drop or two more of the 'good' solvent to clear the opalescence. If in either case the solution becomes opalescent as it cools and the compound starts to separate as an oil, a few more drops of the 'good' solvent should be added and the solution reheated. When an effective procedure has been found it should be followed on the large scale for the crystallization of the bulk sample.

3.2.5 Special topics

(a) The preparation of crystalline samples for combustion analysis

Research and project work often produce new compounds (i.e. ones not previously prepared and reported in the chemical literature). Such compounds are usually identified by elemental analysis and by using the information obtained from various spectroscopic techniques (Chapter 5). These data together with physical characteristics such as melting point (or refractive index for a liquid) are then reported in the literature so that the same compound can be easily recognized again by other workers. Such data are only reliable and reproducible if the compound is absolutely pure. The presence of impurities can produce misleading information leading to great difficulty in structural assignment or at worst the assignment of an incorrect structure to the compound. 'Elemental analysis' is the term commonly used for the determination of the empirical formula of the compound (i.e. $C_xH_yN_z$, etc.) by combustion analysis. It is vitally important not only for the determination of the elemental composition (which can also be done by mass spectrometry) but also as a criterion of purity. For the reasons noted above, data on new compounds cannot be published in reputable chemical journals unless their purity has been demonstrated by correct combustion analysis. This is a vital safeguard for sound science and high professional standards*.

* Exceptionally, some other evidence of purity may be used for compounds that are very unstable and cannot be analysed by combustion analysis.

In combustion analysis a tiny amount of the compound ($c.$ 3–5 mg) is burned in excess oxygen and the amounts of CO_2, H_2O and N_2 are determined. From this the percentages of C, H and N can be determined and calculation will show if these experimental data are consistent with the proposed empirical formula (the acceptable error limits are $\pm 0.3\%$).

Since only a tiny amount of sample is required, it must be not only chemically pure but also free from dust, filter-paper fibres and all other extraneous material. The preparation of samples for analysis must therefore be carried out with great care. All apparatus must be scrupulously cleaned, rinsed with distilled water and oven dried (most conveniently in a large crystal dish covered with a clock glass to keep out dust). During crystallization, hot filtration to remove dust is essential, either by suction filtration (Section 3.2.3) or by the method described on p. 78, which is easier for small samples and minimizes the number of pieces of apparatus involved. A Craig tube (Fig. 3.16) can be used for separating and drying the crystals to minimize handling and exposure to dust, but it is perfectly acceptable to filter off the crystals using a small sintered funnel (Fig. 3.10), or even a Hirsch funnel with a special hard filter paper. However, if the latter method is used, the crystals should be dried by suction on the filter only briefly, to avoid drawing dust into the sample. Finally the crystals should be dried in a drying pistol (Fig. 3.14) to remove occluded solvent. Before submitting the sample to the analyst it should be checked for dust and other contaminents by a careful examination with a hand lens.

(b) Crystallization of compounds of very low solubility

In the trial crystallization it may be found that the compound is too insoluble in all solvents for crystallization by the conventional method. Such compounds can often be crystallized satisfactorily using the continuous Soxhlet extraction procedure, as described in Section 3.1.3. The solid to be crystallized is placed in the porous thimble (Fig. 3.5), and the RB flask filled to $c.$ two-thirds full with an appropriate solvent. This should preferably not be a high-boiling solvent because prolonged reflux at a high temperature could lead to thermal decomposition of the compound. If either the solvent (Section 3.2.4) or the compound are air-sensitive, the procedure should be carried out under nitrogen (Section 2.1.5). Prolonged extraction will transfer the compound gradually into the flask, where it will crystallize.

3.3 MELTING POINT

3.3.1 General principles

The melting point (m.p.) is an important property of an organic solid. It provides a means of identification and a measure of purity. Its use for identification rests on the fact that all compounds, when pure, have a definite melting point, or to be more precise a small temperature range over which they are completely transformed from solid to liquid. This temperature range spans a maximum of 1–2°C for pure compounds. Thus the m.p. of a pure sample of benzoic acid, for example, would be recorded as 120–122°C. The use of m.p. as a measure of purity depends on the fact that the presence of impurity has two effects: it lowers the m.p. and it widens the melting range. For example, a slightly impure sample of benzoic acid might melt at 114–119°C.

The use of m.p. as a means of identification is obviously subject to much uncertainty since there are many millions of organic compounds and, coincidentally, lots of them will have the same melting points. It can be made much more certain, however, by the use of the technique known as 'mixed melting point' determination. This method can only be used if you have an 'authentic' sample of the compound in question. It involves thoroughly grinding together equal amounts of the 'unknown' and 'authentic' samples and taking the m.p. of the mixture. If the mixed m.p. is virtually the same as that of the two samples then it is very likely that they are the same. If the m.p. of the mixture is depressed then it is very likely that they are different.

3.3.2 Determination of melting point

In most routine m.p. determinations a small sample of the substance is placed in a thin-walled capillary tube. The tube (Fig. 3.17) is then heated slowly in a melting point apparatus to determine its melting range.

The most common type of m.p. apparatus, shown schematically in Fig. 3.17, contains an electrically heated metal block which has a hole for the thermometer, three holes for the m.p. tubes and one for observation. The block is mounted in a casing, which shields it from draughts and in which a magnifying lens and a lighting system are mounted so that the samples can be observed closely during heating. The presence of three holes for the m.p. tubes allows the simultaneous investigation of both samples and the mixture in mixed m.p. determinations. The rate of heating of the block is controlled by a rheostat.

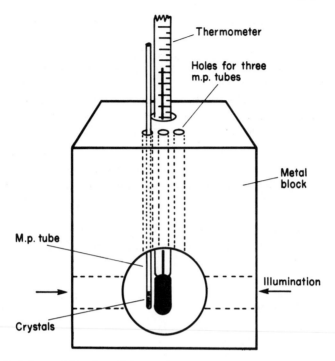

Fig. 3.17 A heated block melting point apparatus.

Some more expensive instruments have a liquid heating bath (for better heat transfer), instead of a metal block; and others have a platinum resistance thermometer with digital temperature readout instead of a thermometer.

The capillary tubes for m.p. determination are usually supplied as open-ended tubes *c.* 10 cm long. Each can be converted into two m.p. tubes by heating the centre in a small hot micro-burner flame and drawing out the tube to make two sealed ends. This requires practice since the ends must be reasonably symmetrical and not bent, or the tube will not fit into the m.p. instrument. Melting point tubes should be kept clean and dust-free in a stoppered tube.

To fill the m.p. tube first crush a little of the sample with a spatula on a clean watch glass and then press the open end of the tube down onto the heap of powder. This will produce a small 'plug' of solid in the open end. It can be persuaded down the tube by tapping the closed end gently on a hard surface or by drawing the edge of a file along the outside of the tube near the top to produce vibration. Finally the tube should be tapped repeatedly on a hard surface (or dropped down a

2 ft length of glass tubing) to produce a compacted column of solid *c.* 2–3 mm long.

If the approximate m.p. of the compound is not known then it saves much time and tedium if a rough determination using fairly rapid heating is carried out first. The instrument is then allowed to cool by *c.* 30°C, a new sample is put in and the heating control adjusted to give a slow temperature increase – about 1°C per minute in the region of the melting point.

Too rapid heating will result in an erroneously low result because the temperature of the thermometer will lag behind that of the block.

Record the temperature range, to the nearest 0.5°C, from the first sign of melting (first appearance of liquid), to the complete disappearance of the solid.

3.3.3 Other methods

(a) The Kofler hot-bench

This is a useful apparatus (Fig. 3.18) for the rapid (*c.* 1 min) determination of routine melting points in the range 50–260°C. It consists of a metal bar heated electrically to maintain a constant temperature gradient along its length. To determine the m.p. of a substance, a few crystals are sprinkled along the bar and the pointer is moved until it coincides with the demarcation line between solid and melt (Fig. 3.18b). The m.p. is then indicated on the temperature scale. The apparatus has an accuracy of *c.* ±1°C but for best results needs to be calibrated in the region of the expected m.p. using one or two of the calibration compounds provided.

(b) The hot-stage microscope

This is a much more advanced instrument, which requires only a few crystals for a m.p. determination. Instruments of this type are normally used for the accurate determination of the m.p. of a new compound. The crystals are set up on a heated/illuminated microscope stage (Fig. 3.19) and can be observed at high magnification as they melt.

The procedure is to place a few crystals of the sample on the glass slide (Fig. 3.19), cover them with a cover slip, and then with the bridge glass and the circular glass cover. The microscope is then focused on

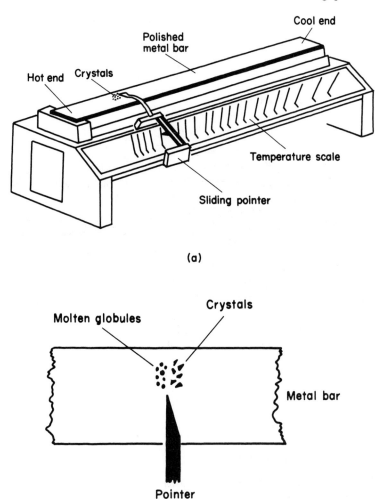

Fig. 3.18 The Kofler hot-bench.

the crystals and the rheostat adjusted to raise the temperature fairly rapidly to about 10°C below the m.p. and then at *c.* 1°C per minute. It will be seen that the crystals lose their sharp edges just before melting and coalesce into small droplets. Record the temperature range from the appearance of fine globules around each crystal to complete melting.

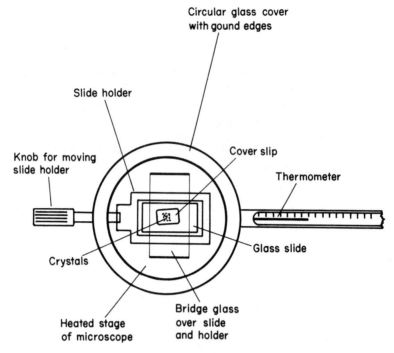

Fig. 3.19 The 'stage' area of a hot-stage microscope.

3.4 DISTILLATION

3.4.1 General considerations

Distillation is the most important and widely used method for the purification of organic liquids and the separation of liquid mixtures. The procedure involves boiling the liquid (distilland) to vaporize it, and then condensing the vapour to give the distillate. The separation of a pair of liquids whose boiling points differ by *c*. 50–70°C or more can be carried out by simple distillation (Section 3.4.2), but if the difference is less, more complicated apparatus is required and the process is known as fractional distillation (Section 3.4.3). Some liquids have boiling points that are too high to allow distillation at atmospheric pressure without causing thermal decomposition. Reducing the pressure lowers the boiling point and thus allows very high-boiling liquids and oils to be distilled easily and safely. The technique is known as vacuum distillation (Section 3.4.5).

SAFETY Great care must always be taken when distilling organic liquids since most of them are flammable. The greatest hazard is posed by those which are also very volatile (e.g. ether, b.p. 35°C). Such compounds must not be distilled in any way that allows their vapour to come into contact with flames, sparking motors, bimetallic power regulators, IR instruments or any other source of ignition. A further hazard is posed by peroxides, which may form when certain compounds, notably ethers and hydrocarbons, are exposed to air (see p. 5). Liquids containing peroxides must NOT be distilled until the peroxides have been removed. Materials provided for use in teaching laboratories should be peroxide-free. However, in project work it is a hazard that must always be borne in mind. A number of tests for peroxides and methods for their removal are given in references 1–3. Reference 3 contains a useful discussion and comparison of the methods available and a recommendation for the use of self-indicating molecular sieve (4A) as an effective reagent for the easy removal of hydroperoxides from common ether solvents such as diethyl ether and tetrahydrofuran.

3.4.2 Simple distillation

(a) At atmospheric pressure

A typical assembly of apparatus for simple distillation is shown in Fig. 3.20a. A system of this type is not very efficient and will not give a clean separation of liquids with a boiling point (b.p.) difference of less than c. 50–70°C. If it is used for a mixture where the b.p.'s are closer, then, although the more volatile component will distil over first, it will be contaminated with the higher-boiling component even in the early stages of the distillation.

APPARATUS. The distillation flask is usually round-bottomed for large volumes and pear-shaped for volumes of 100 ml or less. Select a flask of a size such that it is no more than two-thirds full at the begin-

1. Riddick, J. A. and Burger, W. B. (1970) *Organic Solvents: Physical Properties and Methods of Purification, in Techniques of Chemistry*, 3rd edn, vol. 2, Wiley Interscience, New York.
2. Gordon, A. J. and Ford, R. A. (1972) *The Chemists Companion*, Wiley Interscience, New York.
3. Burfield, D. R. (1982) *J. Org. Chem.*, **47**, 3821.

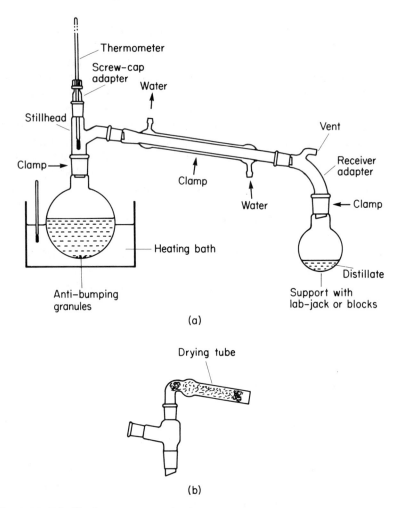

Thermometer

Screw-cap
adapter   Water

Stillhead

Vent

Clamp →

Receiver
adapter

Clamp

Water   ← Clamp

Heating bath

Distillate

Anti-bumping
granules

Support with
lab-jack or blocks

(a)

Drying tube

(b)

Fig. 3.20 Distillation at atmospheric pressure.

ning of the distillation. The flask is connected to the condenser by a
stillhead (Fig. 3.20a). Alternatively, specially designed distillation
flasks (Fig. 3.21a) with an integral side arm are available in the volume
range 10–100 ml. Conical flasks should not be used. When liquids of
high b.p. are being distilled it may be necessary to insulate the stillhead with a wrapping of glass-fibre tape. This will prevent excessive
condensation and thus allow the vapour to reach the side arm.

The thermometer for measuring the distillation temperature is
mounted in the stillhead. To obtain the correct temperature the bulb

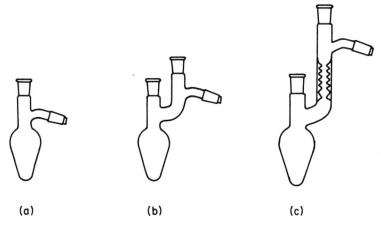

(a)   **(b)**   **(c)**

Fig. 3.21 Distillation flasks: (a) pear-shaped; (b) Claisen; (c) Claisen with Vigreux column.

must be fully in the vapour stream, just below the level of the side arm.

The type of condenser required depends on the boiling point. A simple Liebig condenser (Fig. 2.10) is adequate for liquids of b.p. *c.* 50°C or more, but for very volatile liquids, e.g. ether (b.p. 35°C), a double-surface (Davies) condenser (Fig. 2.10) is necessary. The water flow goes in at the bottom of the condenser and out at the top. A slow flow is usually all that is necessary; excess water pressure will cause the rubber outflow tube to fly about. If the b.p. is *c.* 150°C or above, then an air condenser (Fig. 2.10) will suffice.

The receiver adapter that connects the condenser to the receiver flask must have a vent open to the atmosphere at all times, otherwise pressure will build up in the system with potentially disastrous consequences. If it is essential to exclude moisture, a loosely packed drying tube can be attached to the vent by a short piece of rubber tubing, or the alternative arrangement shown in Fig. 3.20b used.

The distillation flask may be heated with a free Bunsen flame ONLY on the rare occasions when an aqueous solution is being distilled. For organic liquids the distillation flask should be partially immersed in a heating bath (Fig. 3.20a) of some kind. The type of bath and the nature of the heat source depend largely on what is being distilled. An electrically heated water bath can be used for liquids of b.p. up to *c.* 80°C. This is particularly applicable to the distillation of ether or other volatile highly flammable liquids when it is essential that there is no flame or other source of ignition in the vicinity. (On a small scale the alternative is to heat a beaker of water on an electric hot-plate.)

SAFETY Volatile flammable liquids such as ether or light petroleum should always be distilled in a fume-cupboard and not on the open laboratory bench where one of your neighbours may suddenly decide to light a Bunsen burner.

Liquids of b.p. $> 80°C$ require an oil bath. The various types of oil and their temperature limits are discussed in Section 2.1.4 (p. 19). The upper limit for oil baths is c. 250°C using a silicone fluid. Accurate temperature control is vital and the ideal bath is a specially designed electrically heated oil bath with fast response and accurate thermostatic control. However, these are expensive and rarely available in teaching laboratories. Usually the oil is contained in a saucepan, for large-scale work, or an enamelled mug for small-scale work, and heated by a Bunsen (or micro) burner or an electric hot-plate. The latter are generally very slow in response and provide very poor control of the bath temperature. The most common method therefore is to use a Bunsen or micro burner with a screw clip on the rubber tubing to give fine control of the flame size.

SAFETY Flame-heated baths should NEVER be used for the distillation of volatile flammable liquids. Heating mantles are not suitable for heating distillation flasks because of the high surface temperature which can arise as the liquid level falls.

If an oil bath is to be used within c. 50°C of its upper temperature limit it will smoke to some extent and smell unpleasant, and is best set up in a fume-cupboard. If temperatures above 250°C are required then a Woods metal bath (p. 20) can be used up to c. 350°C. However, it is generally better and safer to distil high-boiling point liquids under reduced pressure (see following subsection (b)).

When setting up the apparatus it will be necessary to use several clamps for support. The number can be kept to a minimum by using plastic joint clips (Fig. 2.1) to prevent the receiver adapter and flask from falling off. Once set up, the assembly is not easily raised or lowered so it must be set up at the correct height in the first place. This is determined by the height of the oil (or water) bath, so start assembly at that end with the distillation flask at the correct height.

PROCEDURE FOR DISTILLATION. Set up the apparatus taking account of the notes above. Weigh the receiver flasks. Pour the liquid into the distillation flask, using a funnel to avoid spillage. It is better to do this before the heating bath is put in position so that if any spillage does occur, it does not contaminate the bath fluid. The flask should be no more than two-thirds full at the start of the distillation. Add a few anti-bumping granules to promote smooth boiling. If distillation has to be interrupted, more anti-bumping granules should be added before resuming but NOT when the liquid is at its b.p. or superheated (when it might cause sudden violent boiling and foaming). Check that all joints are tight and the receiver flask is properly supported. Check that the vent tube is open to the atmosphere. If the liquid is flammable, a piece of rubber tubing should be attached to the vent and led to a safe venting place away from any source of ignition (see safety note p. 91). Check that the condenser water is on. If an oil bath is to be used make sure there is no water in it (see p. 20). If all is satisfactory then start heating the bath GENTLY so that the temperature goes up slowly. This allows time for heat to transfer to the liquid in the flask. When it starts to boil, cut down the heating rate and control the bath temperature so that distillation proceeds slowly and steadily. It will usually be found necessary to have the bath at a temperature up to c. 30°C higher than the distillation temperature to produce the required distillation rate. The best separation is achieved by very slow distillation (c. 10 drops per minute). Record the boiling range of each fraction, e.g. b.p. 71–73°C. Do NOT distil to dryness, some residues may contain explosive peroxides (see pp. 5 and 91). After distillation, weigh each fraction.

(b) Distillation under reduced pressure (vacuum distillation)

Many organic compounds cannot be distilled satisfactorily under atmospheric pressure because they undergo partial or complete thermal decomposition at their normal boiling points. Reducing the pressure to less than 30 mmHg* considerably lowers the boiling point and this will usually allow distillation to be carried out without danger of decomposition. Two kinds of vacuum pump are in common use: (i) The water pump (sometimes known as a filter pump or aspirator) will reduce the pressure to 10–20 mmHg. This will lower boiling points by 100–125°C. The operation of water-pump vacuum systems is discussed in Section 3.4.8(a). (ii) The rotary oil pump will give pressures as low as 0.01 mmHg. Below c. 30 mmHg the b.p. is lowered

---

* Pressures are given in mmHg since this is the unit obtained directly from mercury manometers and McLeod gauges (760 mmHg = 1 atm; 1 mmHg = 1 torr = 133.3 N m^{-2}).

by *c.* 10°C each time the pressure is halved. The operation of oil-pump vacuum systems is discussed in Section 3.4.8(b).

At very low pressures (< 0.1 mmHg) the vapour density is too low to allow effective fractional distillation and the process is known as molecular distillation (Section 3.4.5). It provides a useful method of purifying liquids that are very thermolabile or of very high molecular weight.

SAFETY In all vacuum distillations observe the proper safety precautions for working with vacuum systems (Section 1.3.3).

APPARATUS. A typical assembly is shown in Fig. 3.22. It differs from the set-up for atmospheric-pressure distillation in several important respects. The distillation flasks are similar but the stillhead is now of the Claisen type with two necks, one for the thermometer and the other for a capillary leak (see below). For small-scale work (100 ml or less) it is more convenient to use the one-piece Claisen flask (Fig. 3.21b) rather than a separate flask and stillhead. These are available in volumes 10–100 ml. The incorporation of a short Vigreux column (Fig. 3.21c) gives a small but useful increase in efficiency. The flasks should be one-half to two-thirds full at the start of the distillation.

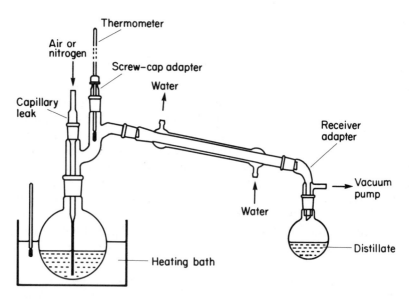

Fig. 3.22 Distillation under reduced pressure.

The capillary leak is a finely drawn capillary tube which reaches down to within 2–3 mm of the bottom of the flask. When the system is under vacuum it admits air (or nitrogen) to produce a very fine stream of bubbles, which promote smooth boiling (anti-bumping granules do not work under vacuum). A low-pressure nitrogen line (*c.* 1 psi) should be connected to the inlet whenever the liquid to be distilled is susceptible to oxidation. A new capillary leak has to be drawn for each distillation. It should be as fine as possible so as to produce the required stream of small bubbles without loss of vacuum. Drawing capillary leaks is a technique better demonstrated than described but the basic procedure is given in Section 3.4.8(c).

The Liebig condenser can be connected to the receiver flask using the simple receiver adapter shown in Fig. 3.22, but this has the major disadvantage that the flask cannot be changed without loss of vacuum and disturbance of the equilibrium and continuity of the distillation. This problem can be avoided by using a more complex system for collection of the distillate. The most common type is the 'receiver adapter with multiple connections' (Fig. 3.23a), commonly known as a 'pig', which allows each flask to be rotated into the receiving position. The 'rotating' joint must be greased. There are many similar designs of rotating collectors, some accommodating up to 12 collecting tubes. The alternative is the Perkin triangle, shown in Fig. 3.23b. Using this device the receiver can be changed by manipulating the taps in the correct order to isolate the flask to be removed, admit air to it, replace it and then evacuate the new flask before interconnecting it to the system.

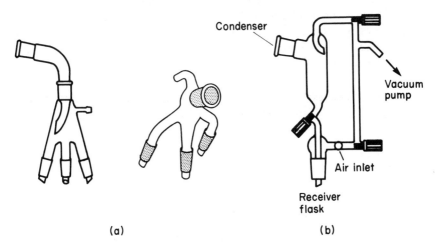

Fig. 3.23 Receiver adapters: (a) multi-arm rotating; (b) Perkin.

The distillation flask should be heated in an oil bath as discussed for atmospheric-pressure distillation (Section 3.4.2).

When setting up the apparatus it is convenient to use plastic joint clips (Fig. 2.1) to attach the flasks to the receiver adapter (pig) and the pig to the condenser, otherwise there is a risk of them falling off before the vacuum is applied. It is not usual to use joint grease for systems at water-pump pressure (10–20 mmHg) (except for the rotating joint of the pig), but it is necessary to use it sparingly for high-vacuum distillation when an oil pump is used.

PROCEDURE FOR VACUUM DISTILLATION.

The liquid to be distilled must be free from volatile solvents such as ether, otherwise the sudden drop in pressure when the vacuum is applied will cause an uncontrollable frothing of the contents of the flask up through the stillhead and into the condenser.

Set up the apparatus taking account of the notes above; place the liquid in the flask (not more than two-thirds full) but do not fit the capillary leak at this stage (or the capillary will fill with liquid) and do not heat the flask before the vacuum is applied. Check over the apparatus. Check the vacuum system to be used before connecting it to the distillation assembly to make sure it will deliver the vacuum required. When all is satisfactory, fit the capillary leak, connect the vacuum system and evacuate the distillation assembly slowly. When the desired pressure has been reached, you can start heating the oil bath. When collecting the distillate note the pressure and the temperature range for each fraction. After the distillation is finished, remove the source of heat and allow the apparatus to cool before releasing the vacuum.

This is not released by turning off the pump but by opening an air-inlet tap in the system (see Section 3.4.8, pp. 111 and 112).

Take precautions to ensure that the receiver flasks do not fall off when the vacuum is released.

3.4.3 Fractional distillation

The efficiency of separation can be improved markedly by the use of a fractionating column. At best this will allow the separation of liquids whose boiling points differ by only a few degrees Celsius. The fractionating column (e.g. Fig. 3.24) consists of a vertical tube mounted above the distillation flask, containing packing material of high surface area onto which partial condensation of the vapour takes place. The objective is to achieve the most intimate contact between the ascending vapour and the descending condensate, which flows continuously down the column and back into the distillation pot. Under these conditions the column will reach a state of equilibrium in which the vapour emerging at the top consists ideally of the more volatile

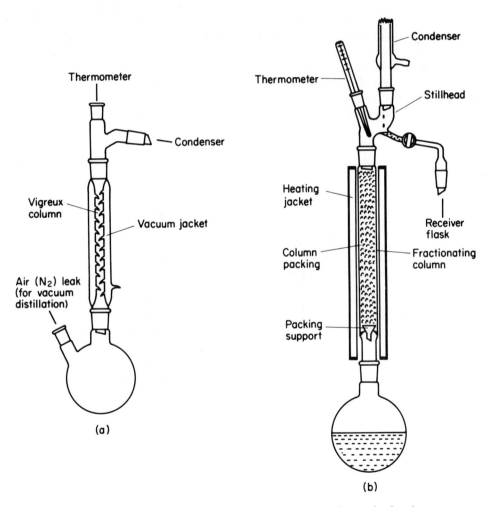

Fig. 3.24 Fractional distillation: (a) Vigreux column; (b) packed column.

component only. This process is known as contact rectification and its efficiency depends on maximizing the surface area of the descending liquid within the fractionating column.

The design of fractionating columns and their operation is a complex and interesting area, which can be treated here only briefly.

(a) Column types

Fractionating columns fall into several different general types.

The first is those in which the high surface area is achieved by filling a tubular column (Fig. 3.24b) with loose packing material, e.g. Raschig rings (short lengths of glass tubing), Fenske helices (small single- or double-turn helices of glass or metal) or stainless-steel gauze rings. These are sometimes called 'dump' packings and columns of this type are generally referred to as packed columns.

The second contains structured packings made of 'knitted' or preformed stainless-steel gauze.

The third type is represented by columns of the thin-film type in

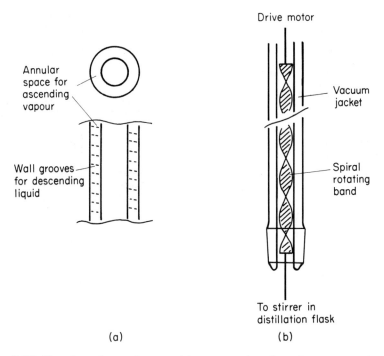

Fig. 3.25 Fractionating columns: (a) concentric tube; (b) spinning band.

which the column interior is relatively open but has an internal structure that forces the vapour to pass close to glass surfaces down which the condensate is flowing. These include the Vigreux column, a simple glass tube with downward-pointing indentations (Fig. 3.24a). The most efficient system of this type is the concentric-tube column (Fig. 3.25a) in which the vapour ascends through a narrow annular space between two walls down which the condensate flows, guided by spiral grooves. A range of such columns – 'Spaltrohr' columns – is commercially available*.

Finally, there are spinning-band columns (Fig. 3.25b) in which the condensate and vapour are mixed by a rapidly spinning spiral band.

(b) Column properties

Three properties of the columns are of interest.

(i) Efficiency

The efficiency of a column is measured in theoretical plates (one theoretical plate is the efficiency of the simplest 'simple' distillation). A useful means of comparing columns of different types is the 'height equivalent to a theoretical plate' (HETP). Packed columns (helices and gauze rings), concentric-tube columns and spinning-band columns all have high efficiency with HETPs in the range 0.5–2 cm. The Vigreux column is much less efficient and has an HETP of *c.* 15 cm at best.

(ii) Hold-up

This is the volume of liquid required to cover the inside of the column and its packing. Since it cannot be recovered it represents material which is 'lost' in the distillation. Packed columns have high hold-up and are thus not suitable for the distillation of small volumes of material. Vigreux, concentric tube and spinning band columns have low hold-up.

(iii) Pressure drop

This is important in vacuum distillation and is a measure of the pressure required in the flask to drive the vapour through the column into the stillhead. Obviously the lower the pressure drop the better the vacuum in the flask and the lower the boiling point. This may be important in the distillation of heat-sensitive compounds. Packed

---

* Fisher Spaltrohr Distillation System, marketed in the UK by Orme Scientific, Middleton, Manchester.

columns have a high pressure drop, while Vigreux, concentric-tube and spinning-band columns do not.

In operation the heat loss from the fractionating column is minimized either by a vacuum jacket (sometimes silvered) (e.g. Fig. 3.24a) or an electrically heated jacket (Fig. 3.24b).

(c) Stillheads

A considerable improvement over simple distillation can be achieved by simply interposing a Vigreux or a short packed column between the flask and a normal stillhead (Fig. 3.24a). This arrangement can be used for atmospheric-pressure or vacuum distillation. In such a set-up the best separation is achieved by very slow distillation.

When really high efficiency is required the normal stillhead is replaced by a total reflux partial take-off type (Fig. 3.24b), which allows the operator to control the reflux ratio (the ratio of the amount of liquid taken off as distillate to that which is returned down the column). In general, the slower the take-off, the higher the purity of the material collected. A more complex stillhead with the facility for multiple fraction collecting is required for vacuum fractionation.

OPERATION. If a column heating jacket is being used it is usual to keep the temperature at the top *c.* 5°C below the distillation temperature. The system is initially operated under total reflux, first to wet the packing material thoroughly, and then to establish the primary equilibrium. This may take some time (several hours) and is indicated by the head temperature reaching a minimum value. A more direct indication is by GLC (Section 4.1.2) and some stillheads have septum-sealed sampling ports so that the composition of the head condensate can be regularly monitored. Collection is then started using a reflux ratio that will give the required separation – again GLC is enormously helpful in establishing an appropriate take-off rate.

3.4.4 Small-scale distillation

The purification of small amounts of liquid (1–10 ml) by distillation is difficult and can involve severe loss of material unless the apparatus is chosen with care. The major problem is the high percentage loss of material caused by hold-up, i.e. the unrecoverable material that forms a film over the surface of the flask, condenser and other glassware. This can be minimized by using very small apparatus designed to have a minimum wetted area. All small distillation flasks are pear-shaped to minimize thermal decomposition.

Amounts in the 10–1 ml range can be distilled using apparatus of the type shown in Fig. 3.26 in which the cold-finger condenser is integrated into the receiver adapter. For amounts at the top end of the range the flask can be of the type shown in Fig. 3.26a, with a short Vigreux column to give some fractionation, and fitted with a capillary leak. However, for smaller amounts the simple pear-shaped flask (Fig. 3.26b) is preferable. It is impractical to use a capillary leak in these very small flasks and bumping is prevented by loosely packing the lower part of the bulb with glass wool, or by using pumice powder as an anti-bumping agent. Flask sizes down to *c.* 2 ml can be used.

Very small volumes (*c.* 1 ml or less) can be distilled using the apparatus shown in Fig. 3.27. The liquid to be distilled is placed in the bottom of the outer tube and heated in a small oil bath. The distillate condenses on the cold finger and collects in the glass bucket. Bumping can be prevented by using glass wool as discussed above. An alter-

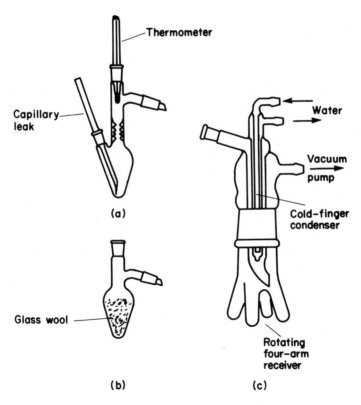

Fig. 3.26 Small-scale distillation.

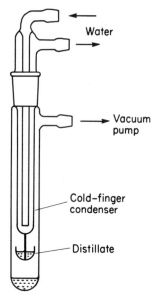

Fig. 3.27 Very small-scale distillation.

native method is to carry out the distillation using a sublimation tube with a short side arm to act as a receiver (see Fig. 3.32b). The liquid can be put into the tube using a long Pasteur pipette, or better, in a sample vial, as for sublimation (Section 3.4.7). The side arm is cut off after distillation.

Another approach to reducing hold-up is to eliminate the condenser altogether and to distil directly into a cooled receiver (Fig. 3.28). This works well but has the disadvantage that the receiver cannot be changed without loss of vacuum. An extension and major improvement on this principle is found in the popular Büchi Kugelrohr distillation system, shown schematically in Fig. 3.29. It can be used for amounts in the 20 mg – 20 g range and allows rapid distillation with minimum heating, so making it particularly useful for thermolabile compounds and high-boiling oils. However, it does not give very effective fractionation and is best for mixtures with a b.p. difference of at least 20–30°C. In this system the distillation is carried out in a set of ball tubes (Fig. 3.29) with interconnecting ground-glass joints. This assembly is mounted on a sliding strut so that any number of bulbs can be inserted into the oven. In the more expensive versions the tubes can be rotated by an electric motor (like the flask on a rotary evaporator). The oven has precise temperature control and is made of metal on the simpler version and glass (covered with a

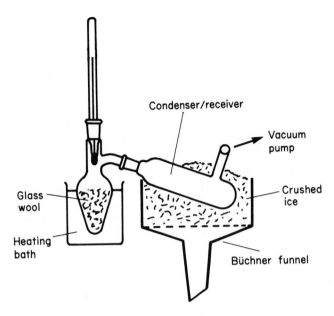

Fig. 3.28 Small-scale distillation into a cooled receiver.

metallic heating film) on the more expensive type. The entrance to the oven is fitted with an iris shutter, which can be closed around the connectors between the bulbs of the ball tube.

The liquid to be distilled is placed in the end tube (or in a standard 25 or 50 ml B14 RB flask if a larger volume is required). Very small

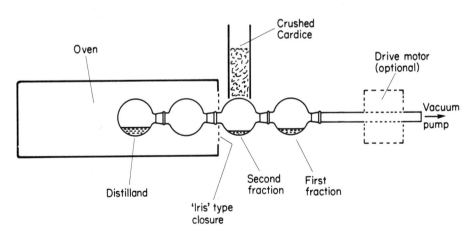

Fig. 3.29 Ball-tube distillation (Kugelrohr).

amounts of involatile liquids can be put in as solutions in ether or dichloromethane and the solvent then removed using a rotary evaporator. The bulb can be loosely packed with glass wool to prevent bumping, but this reduces the speed of distillation considerably and is not advisable with thermolabile materials. It is not necessary with the 'rotary' Kugelrohr.

For simple distillation, when only one fraction is expected, the end bulb only is inserted into the oven and the iris closed to shield the outer bulbs from the heat. The vacuum is then applied – gently to avoid foaming – and the rotation (if available) is started. The oven temperature is then raised until distillation commences. For liquids of low and moderate b.p. it is necessary to cool the 'condenser' bulb. This can be done conveniently by clamping a wide-bore tube just above the bulb (Fig. 3.29) and filling it with powdered Cardice (Section 2.1.4). For fractional distillation, all the bulbs are inserted into the oven except the outer one, which serves as the receiver for the first fraction. The bulbs are then withdrawn to collect successive fractions as required.

Fractional distillation of small volumes is again difficult because of hold-up losses, which effectively rule out the use of packed columns (Section 3.4.3). However, substantial improvements can be achieved using thin-film type columns (Section 3.4.3) in which the problem is less severe. Some increase in efficiency is obtained by the incorporation of a Vigreux column (Fig. 3.26), preferably vacuum-jacketed to minimize heat loss. Very high efficiency can be obtained (at much greater expense) by using the micro Spaltrohr system (Section 3.4.3). This is suitable for amounts in the $1-10$ ml range, has an efficiency of 15 plates and a hold-up of 0.1 ml. Very small amounts (≤ 1 ml) cannot be fractionated with efficiency and must be separated by preparative gas–liquid chromatography, if reasonably volatile, or by preparative TLC or one of the forms of column chromatography, if not.

3.4.5 Molecular distillation

Normal vacuum distillation at pressures down to 10^{-1} mmHg is not practicable for some liquids either because they have very high boiling points or because of thermal instability. In many cases such materials can be purified at very low pressure ($10^{-3} - 10^{-6}$ mmHg) by molecular distillation in a still designed so that the gap between the liquid surface and the condenser is less than the mean free path of the molecules. The simplest version of such a still is shown in Fig. 3.30a. In molecular distillation the molecules proceed directly from

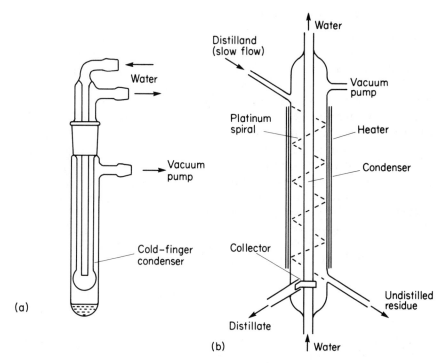

Fig. 3.30 (a) Small-scale molecular distillation or sublimation; (b) falling-film molecular still.

the liquid surface to the condenser without having to pass through a barrier of air molecules and few therefore return to the liquid. The normal concept of 'boiling point' does not apply since there is no longer an equilibrium between liquid and vapour.

Very small quantities can be distilled using the simple apparatus in Fig. 3.30a or the 'bucket' variation in Fig. 3.27. Larger amounts are processed using a 'wiped-film' or 'falling-film' molecular still (the latter shown schematically in Fig. 3.30b) in which the distilland is spread in a thin film over the inner wall of a tubular heater and distils across to the cooled tube in the centre. Bulb-to-bulb distillation as in the Kugelrohr system (Fig. 3.29) can also be carried out under high-vacuum conditions but the ultimate vacuum is limited in the 'rotating bulb' version by the vacuum seal.

The high vacuum required can be produced by a diffusion pump backed by a rotary oil pump. In small-scale distillations the vacuum must be applied very gently to avoid frothing and the bath temperature raised very slowly to avoid bumping.

3.4.6 Steam distillation

Some organic compounds that are virtually immiscible with water may be separated from non-volatile impurities including inorganic contaminants by steam distillation. This process is essentially a co-distillation with water and is usually accomplished by passing a current of steam through a hot mixture of the material to be distilled and water. Provided that the compound possesses an appreciable vapour pressure (5 mmHg or more at 100°C) it will be carried over with the steam and, being immiscible, can be readily separated from the distillate. One of the advantages of steam distillation is that the temperature never exceeds the boiling point of water. This permits the purification of high-boiling substances that are too heat-sensitive to withstand ordinary distillation. The method is also of importance in the separation of volatile products from tarry material, which is often produced during the course of an organic reaction and cannot be removed easily by distillation or crystallization.

A typical apparatus used for the steam distillation of large quantities of material is shown in Fig. 3.31. The material to be steam-distilled is placed in a 500 ml round-bottomed flask fitted with a steam splash-head. This form of head minimizes the chance of non-volatile material being carried over into the condenser by frothing or splashing. The vapour outlet is connected to an efficient condenser and the distillate is collected in a large flask of appropriate capacity.

Steam may be produced in a generator (CAUTION) or taken from a steam line. In the latter case, a trap is usually necessary to allow re-

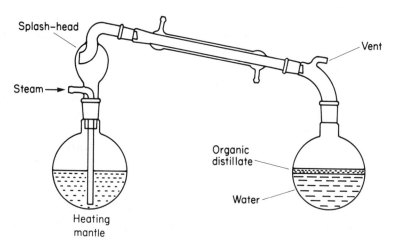

Fig. 3.31 Steam distillation.

moval of condensate before the steam is passed into the distillation flask. It is usual practice to conduct the distillation as rapidly as possible without exceeding the capacity of the condensing system. Since some of the steam necessarily condenses in the distillation flask, it should be heated during the distillation on a heating mantle so that it is kept about half full. If the material being distilled is a low-melting solid, care must be taken that the condenser does not become choked up. Any material that accumulates in the condenser may be quickly cleared by draining the water from the jacket for a few minutes and allowing the steam to carry the melted solid into the receiver.

For steam distillation on a small scale it is often convenient to generate the steam *in situ* by boiling an aqueous solution or suspension of the organic compound. Additional hot water may have to be added during the distillation to prevent the solution becoming too concentrated. This can be achieved without dismantling the apparatus by having a dropping funnel fitted to a side arm of the distillation flask.

3.4.7 Sublimation of solids

Sublimation is one of the most convenient methods for the purification of small amounts of organic solids. In this process the substance is volatilized by heating at a temperature below its melting point (m.p.) and the vapour condensed directly into the solid state on a cold receiver. Only those organic compounds with relatively high vapour pressure at temperatures below their m.p. can be sublimed at atmospheric pressure. These are comparatively few and the majority may be sublimed only under greatly reduced pressure. The technique works particularly well for non-polar compounds, which are generally more volatile than polar ones of similar molecular weight. The method has the advantages of simplicity, ease of operation and minimal loss of material. It is particularly valuable for compounds that solvate or deliquesce. The main disadvantage is that, like distillation, the separation process depends on difference in vapour pressure and therefore compounds of similar volatility will co-sublime. A typical all-glass sublimator is shown in Fig. 3.30a (which can also be used for molecular distillation (Section 3.4.5)). Various sizes are available for dealing with quantities from a few milligrams up to *c.* 1 g.

PROCEDURE. The substance to be sublimed is finely pulverized and placed in the bottom of the outer tube into which is fitted the cold-finger condenser. The distance between the bottom of the sublimator and the tip of the condenser should be short, but sufficient to avoid contamination of the sublimate by spattering of the solid, a problem that is frequently encountered if solvents and other volatile materials are not removed prior to sublimation. A distance of about 1 cm appears to be satisfactory for most cases.

After evacuation with a water or oil pump, the sublimator is immersed in a shallow oil bath and is heated gradually until a film of sublimate collects on the surface of the cold finger. The temperature should not be raised too rapidly or the substance may spatter. When the sublimation is complete, the vacuum is released cautiously and the cold finger is withdrawn. The sublimed material is then detached with a micro-spatula onto a watch glass or a hardened filter paper, or by washing it off with a solvent (although this loses the advantage of direct production of the pure solid).

Very small quantities are better sublimed in a long glass tube (c. 9 mm diameter) heated in an inclined electrically heated metal block (Fig. 3.32a). The impure material is put into a small vial and this is placed at the bottom of the tube. A small plug of cottonwool near the top of the tube will keep out any debris from the rubber tubing. The tube is angled at c. 45°C and then evacuated slowly to avoid sample dust being drawn up the tube. Very volatile solids tend to condense over a considerable length of the tube and it is sometimes necessary to wrap a cooling coil or a piece of wet cottonwool around the tube to promote localized condensation. To remove the sublimate the vacuum is released very carefully and the tubing on either side of the

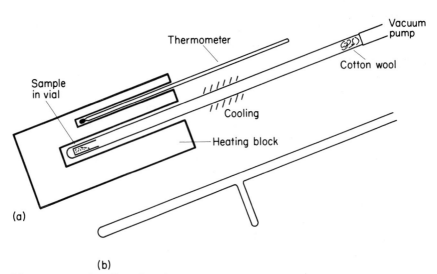

Fig. 3.32 (a) Small-scale sublimation using a heating block; (b) small-scale distillation tube.

sublimate is cut off. The sublimate can then be detached by scraping with a micro-spatula.

3.4.8 Appendices

(a) Water-pump vacuum systems

The water pump (aspirator) provides an easy way of generating a moderate vacuum (10–25 mmHg). The lower limit depends on the water vapour pressure and hence on its temperature, but it should generally be possible to reach 10–12 mmHg with a normal cold-water supply. The usual components of the system are shown in Fig. 3.33. All rubber tubing must be heavy-wall 'pressure' tubing. The manometer is essential for distillation and generally useful to give an indication of the performance of the system. The pressure is read as the difference between the two mercury levels. In operation the water tap is always kept fully turned on; it is not possible to control the vacuum by adjusting the water flow rate. One problem with these pumps is that the water may 'suck back' into the apparatus being evacuated if the water pressure drops. Damage and contamination

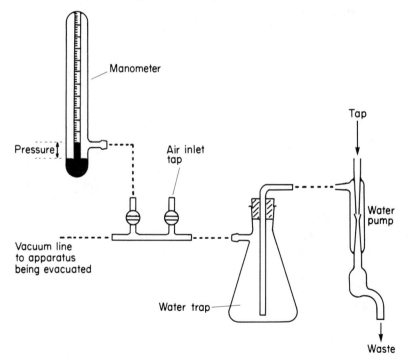

Fig. 3.33 Water-pump vacuum system.

are minimized by fitting a water trap as shown. For the same reason the water tap should not be turned off until the vacuum has been released by opening the air inlet tap.

(b) Rotary oil pumps

Rotary oil pumps – often available as mobile 'trolley pumps' in teaching laboratories – will provide high vacuum for distillation and drying purposes down to *c*. 0.01 mmHg. The system (Fig. 3.34) must incorporate a cold trap to prevent contamination of the pump oil by solvent vapours. The vacuum system should always be set up and tested before connection to external apparatus. The procedure is as follows. Make sure the cold trap is empty and its joint is lightly greased with silicone grease. Immerse the trap slowly in a Dewar vessel of liquid nitrogen, close taps A and B to isolate the system and then switch on the pump immediately.

SAFETY Do not leave the trap immersed in liquid nitrogen unless the system is under vacuum, or liquid oxygen will condense in the trap. Mixtures of flammable solvents and liquid oxygen can cause violent explosions.

Check the vacuum. If it is not *c*. 1 mmHg or better, suspect a leak. The most common cause of leaks is channelling of the grease in the taps or perished rubber tubing. If there are no leaks the problem may be due to the presence of solvent vapours in the pump oil. These can

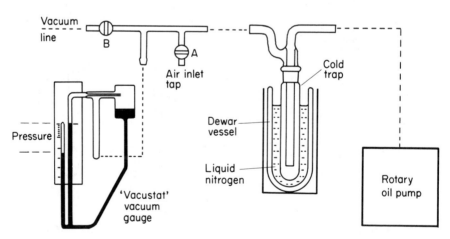

Fig. 3.34 Rotary oil-pump vacuum system.

be removed by opening the air ballast valve for a while to blow out the solvents. The vacuum will be poor while ballasting but should improve markedly when the valve is closed again.

When a satisfactory vacuum has been obtained connect up the external system and evacuate it slowly. To close the system down after use, first isolate the system that has been evacuated, then open tap A to admit air, turn off the pump and remove the trap from the liquid nitrogen. Never close down by simply turning off the pump, leaving the trap under vacuum, or pump oil will 'suck back' into the trap.

(c) Air (capillary) leaks

An air leak for vacuum distillation is drawn from a piece of tubing as shown in Fig. 3.35. The oxy/gas flame should be hot enough to soften the tubing but not so hot that it becomes unworkably fluid. The procedure described is the 'double-drawing' method. First rotate the tubing slowly in the flame until it softens; keep rotating until the walls thicken and then draw the two ends slowly apart to form the first, thick-walled, constriction (Fig. 3.35b). Allow it to cool briefly and then heat the constriction in the tip of the flame until it softens and then draw the ends apart, again with careful control to form the second, very fine, constriction (Fig. 3.35c). Pinch off the capillary to the required length. Allow it to cool and test by immersing the end in a solvent (not near the flame) and blowing gently. A good air leak will give only a very fine stream of bubbles.

Do not work closer than *c.* 1 inch to the ground-glass joint or it will become distorted; at that stage the air leak should be returned to the glassblower to have another length of tubing attached.

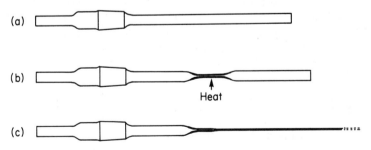

Fig. 3.35 Drawing an air (capillary) leak for vacuum distillation: (a) air-leak tube; (b) first constriction; (c) after second drawing.

4 Separation of organic mixtures by chromatography

'Chromatography' encompasses a range of very powerful experimental techniques for separating mixtures of organic compounds. In this chapter these techniques have been divided into two groups: (a) the small-scale 'analytical' methods used for the qualitative and quantitative analysis of mixtures, and (b) the larger-scale 'preparative' methods used for carrying out separations on the milligram to multigram scale.

Expertise in these techniques is usually built up gradually in teaching laboratories over a number of years. In the early stages the emphasis is on learning to handle the equipment and, at the beginning, the operating parameters to be used (nature of adsorbents and eluting solvents, column packings in gas–liquid chromatography, etc.) will be specified precisely in the instructions. As progress is made the practitioner will acquire an increasing understanding of how these variables affect the separation process and will learn how to optimize them for various kinds of mixtures and reaction products. The structure of each of the sections below is intended to complement this progressive familiarization process. The early parts deal with the straightforward manipulative operations and the later parts with optimizing the separation parameters. For the latter it is important to understand the general basic principles underlying the separation processes and these are outlined in a separate section (Section 4.3.1, p. 170), which should be read as appropriate.

4.1 ANALYTICAL METHODS

4.1.1 Thin-layer chromatography

Thin-layer chromatography (TLC) is one of the most widely used forms of chromatography, and is of enormous value for quick qualitative analysis of mixtures, for monitoring reactions and for determining the operating parameters to be used in preparative-scale column chromatography. Preparative TLC is dealt with in Section 4.2.1.

(a) General description

The separation is carried out on a flat plate coated with a thin layer of an adsorbent such as silica gel or alumina (Fig. 4.1a). The mixture to be separated, dissolved in an appropriate solvent, is spotted onto the plate (Fig. 4.1b) and after the spotting solvent has evaporated the plate is placed in a developing jar (Fig. 4.2c) containing a little of the developing solvent. The solvent rises through the adsorbent layer by capillary attraction, and the various compounds in the mixture ascend at different rates depending on their differing affinity for the absorbent. When the solvent has almost reached the top of the adsorbent layer the compounds should ideally be well separated (Fig. 4.1d). The separation process is one form of liquid−solid chromatography, which is discussed in more detail in Section 4.3.1.

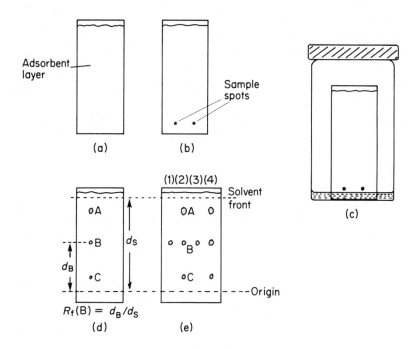

Fig. 4.1 Thin-layer chromatography.

(b) TLC plates and adsorbents

The plates consist of some sort of backing material, which is usually glass but may be plastic or heavy aluminium foil, on which is coated a thin layer of the adsorbent *c*. 0.25 mm thick (Fig. 4.1a). The adsorbent is usually either silica or alumina of small particle size (*c*. 10–30 µm) and contains a binder (*c*. 10% gypsum or starch) to ensure cohesion of the layer. The plates are of two general types: re-usable and disposable.

Re-usable plates are made of thick glass and can be coated with adsorbent using special spreading equipment and then cleaned off again after use ready for recoating. They come in various sizes from 20 × 5 cm down to plates made from microscope slides (75 × 25 mm), which are cheap and useful for monitoring reactions. In teaching laboratories the plates will normally be supplied ready-coated, but for more advanced work details of the spreading process are given in Section 4.3.3.

Disposable plates are supplied by various manufacturers, ready-coated with a robust layer of adsorbent on a backing of either thin glass, plastic sheet or aluminium foil. The last two are particularly convenient as the large sheets can easily be cut up with scissors into economic sizes for use. They give better resolving power than re-usable plates and for most separations a plate height of 5 cm is enough. When using them it is convenient to cut a 5 × 20 cm strip, mark on a light pencil line about 5 mm from the long edge to indicate where to apply the sample spots, and then cut off pieces of appropriate size (e.g. Fig. 4.19, p. 160).

Plates are prepared or supplied to a particular activity (Section 4.3.2), usually II to III, and should be stored over silica gel.

(c) Application of the sample

The sample is normally applied as a 1–2% solution from a capillary dropper. The solution should be made up in a volatile solvent such as dichloromethane or ether (avoid polar solvents such as ethanol), most conveniently in a small sample vial. The dropper is easily made from a melting point tube by carefully drawing out the middle over a microburner flame and then breaking it at the constriction. The position at which to apply the spots depends on the size of the plate – on large (5 × 20 cm) plates they should be about 1 cm from the bottom and on the small disposable plates about 5 mm. The important point is that you leave enough space at the bottom (Fig. 4.1b and c) so that the spots are not immersed in the developing solvent.

To apply the sample first charge the dropper by dipping the end into the solution, when it will fill by capillary attraction. Then just touch the tip onto the adsorbent surface. Try to avoid breaking the surface since this may cause some distortion of the spot when the chromatogram is being developed. It is important to keep the spot as small as possible and this requires the use of a fine capillary. Allow the spotting solvent to dry completely before developing the chromatogram.

(d) Running (developing) the chromatogram

SAFETY This operation usually involves the use of volatile flammable solvents and must be carried out away from sources of ignition, preferably in a fume-cupboard.

This is done by immersing the lower edge of the plate in the developing solvent contained in a tank or jar of appropriate size (Fig. 4.1c). Special jars are available for the large plates (20 × 5 cm), and wide-necked screw-cap bottles are convenient for small ones. Guidance on the choice of developing solvent is given in subsection (g) below. The jar should be partially lined with a piece of filter paper that dips into the solvent; this keeps the atmosphere saturated with solvent vapour and minimizes evaporation from the plate. To run (develop) the chromatogram, first adjust the solvent depth so that the sample spots will be above the surface and then place the plate in the jar leaning on the wall, as near to vertical as possible. This is easy with large plates but the small light disposable plates need to be handled with great care using tongs or tweezers when putting them into the jar and removing them. The jar should be capped during development. When the solvent front is close to the top of the plate, remove it from the jar and immediately mark the position of the solvent front with a pencil or a spatula. Allow the solvent to evaporate from the plate – in a fume-cupboard.

(e) Examining the chromatogram

If the compounds in the sample are coloured there will obviously be no problem in seeing them after development, but for colourless compounds some visualization technique is required.

The most common method is to incorporate an inorganic fluorescent agent (c. 0.5%) into the adsorbent layer. Disposable plates can be bought in this form. When illuminated with an ultraviolet (UV) lamp (254 nm) the adsorbent then glows a pale green or blue colour and the organic compounds show up as dark spots because they quench the fluorescence.

SAFETY The special UV lamps for TLC are of low power but they should be mounted in a viewing box or hood to keep the UV light from the eyes of the operator and to keep out daylight.

Another general method is to use an 'iodine jar'. This is a container of similar size to the developing jar containing a few crystals of iodine. When the dry developed plate is allowed to stand in the jar for a few minutes the iodine vapour dissolves in the organic 'spots' and stains them brown. Eventually the whole plate will darken.

Whatever method is used for visualization, the position of the spots should be marked with a pencil or spatula for later measurement (Fig. 4.1d).

Note that although the above methods are generally effective, some compounds 'mark' more strongly than others and, rarely, some do not show up at all. Do not therefore be tempted to take the relative intensities of the spots as even a rough indication of the relative concentrations in the mixture.

A variety of spray reagents can be used to colour particular classes of compound.[1,2]

(f) The use of TLC for qualitative analysis

Under a given set of conditions (adsorbent and solvent) the R_f value (Fig. 4.1d) of a compound is characteristic. Thus identity of the R_f value of a compound in a mixture with that of an 'authentic' sample of the compound provides a good indication that they are the same (see the caution below). Since adsorbents vary and solvent mixtures are difficult to reproduce accurately, it is necessary to demonstrate that the R_f values are the same by running the mixture and the authentic sample side by side on the same plate and to make sure by adding a little of the 'authentic' to a separate sample of the mixture to see whether the spots exactly coincide. See Fig. 4.1e where (1) and (3) are the 'authentic' compound B, (2) is the mixture and (4) is the mixture plus 'authentic'.

1. *TLC Visualisation Reagents and Chromatographic Solvents*, Eastman Organic Chemicals, Rochester, New York (Kodak Publication No. JJ5).
2. *Dyeing Reagents for Thin-Layer and Paper Chromatography*, E. Merck AG, Darmstadt, Germany.

CAUTION Identity of chromatographic behaviour even on different adsorbents with different solvents can never be regarded as absolute proof of structural identity. The compound concerned must be separated out by preparative chromatography and a further point of identity established (e.g. IR or NMR spectra).

(g) Choosing the developing solvent (mobile phase)

The distance moved up the plate by a compound 'spot' depends on its affinity for the adsorbent and the strength (polarity) of the developing solvent (see Table 4.5, p. 172). Polar compounds (alcohols, ketones, etc.) are strongly adsorbed and may move hardly at all with weak solvents such as hexane, whereas a non-polar hydrocarbon such as naphthalene would be moved well up the plate. The best solvent for a particular mixture has to be found by trial and error. Solvent strength is most easily varied by using mixtures of a strong and a weak solvent. A common procedure is to start off by using a 1:1 mixture of ether (strong) with hexane (or 60/80 petroleum) (weak), and then varying the ratio as appropriate. Obviously if pure ether is not strong enough to move the spots well up the plate, you move to a stronger 'strong' solvent (Table 4.5).

In cases where separation is difficult, i.e. there is a substantial overlap between two spots even when they are pushed well up the plate (R_f 0.6–0.9), you can try to improve matters in two ways: (i) by changing the chemical nature of the developing solvent but keeping the strength about the same (this will change the k' values and hence α; see p. 170), or (ii) by changing the nature of the stationary phase, i.e. alumina rather than silica or vice versa. In practice the latter is often more effective, and quicker if the plates are available.

In cases where a mixture contains several polar and several non-polar compounds you may need to run it twice, first in a weak solvent to give a good separation of the non-polar compounds (with the polar ones left at the origin) and then with a stronger solvent to separate the polar compounds (with the non-polar ones bunched together at the solvent front).

4.1.2 Gas–liquid chromatography

Gas–liquid chromatography (GLC) is a technique used for separating mixtures of 'volatile' compounds, which in practice means those of molecular weight up to *c.* 500. It is of enormous value for both qualitative and quantitative analysis and, because of its very high sensitivity, it can be used to detect the presence of compounds down to the microgram (10^{-6} g) level.

(a) General description

The essentials of the gas chromatograph instrument are shown in Fig. 4.2a. The actual separation of the mixture takes place on the column, which contains the stationary phase – a thermally stable involatile liquid – supported on an inert granulated solid. Silicone oils, hydrocarbon greases and high-molecular-weight polyesters are commonly used as stationary phases. The mobile phase is a gas – usually nitrogen – which is passed continuously through the column throughout the operation. It is generally known as the carrier gas. The column is mounted in an oven, which can be set at temperatures up to 300°C. At one end of the column is attached an inlet for the carrier gas and an injection head, which contains a rubber septum through which the sample can be injected by syringe.

A small volume of the mixture to be separated is injected into the column where it is vaporized, sometimes with the help of a flash heater. The components of the mixture partition themselves between the stationary liquid phase and the mobile phase and move through the column, ideally at different rates, separating into discrete bands which are eluted one after the other through the detector. This device senses the presence of an organic compound in the gas stream and sends an electrical signal to a chart recorder or a visual display unit (VDU). Each compound in the mixture therefore produces a peak on the recorder trace, e.g. Figs. 4.2a and 4.5. In general, the more volatile a compound is, the faster it will pass through the column. This separation process is discussed in more detail in Section 4.3.1.

(b) The instrument

In the early use of GLC in the teaching laboratory the instrument will normally be set up in advance with the correct column, oven temperature and gas flow rate for the particular experiment. A selection of these parameters is discussed in later sections.

Most GLC instruments are fitted with a flame ionization detector (FID) (Fig. 4.2b). In this the gas stream from the column is mixed with hydrogen in a *c.* 1:1 ratio and the mixture is burned at an insulated jet inside the body of the detector. Air is also supplied to support the flame and to sweep away the combustion products. The jet tip forms one of a pair of electrodes, the other being a cylinder surrounding the flame. When one of the eluted organic compounds burns

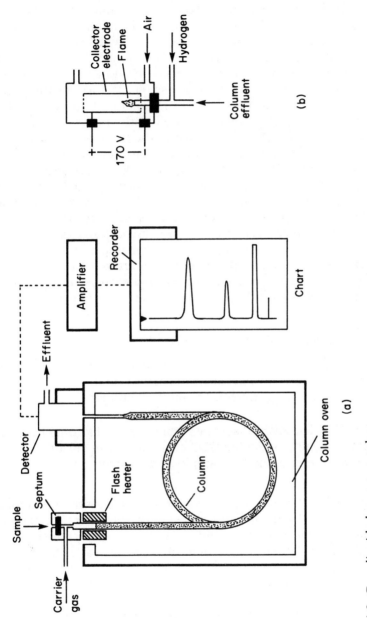

Fig. 4.2 Gas–liquid chromatography.

in the flame it is ionized; the charged particles increase the electrical conductivity between the two electrodes and a minute current (10^{-12} to 10^{-7} A) flows across the gap. This is amplified to give a signal, which is fed to the chart recorder. Flame ionization detectors are popular because of their very high sensitivity (down to 10^{-12} g ml^{-1}) and because they give a linear response, i.e. the size of the peak is directly proportional to the amount of compound burned.

Some gas chromatographs are fitted with thermal conductivity detectors, which monitor the change in the thermal conductivity of the gas stream when an organic compound is present. They are much less sensitive than FIDs and are rare on modern instruments.

The amplifier is fitted with an attenuator switch, which controls the magnitude of the output signal (like the volume control on a radio). This is used to adjust the size of the peaks either to bring them 'on scale' if they are very large or to increase the peak size of minor components so that they are big enough to measure accurately. The attenuator will be calibrated over the range from ×1 (most sensitive position) to ×500 000 (least sensitive), and each peak on the chart should be labelled with the attenuation used to produce it.

(c) Injecting the sample

Samples are usually injected as solutions in a volatile solvent. A concentration of c. 1% is convenient but the technique can cope with solutions that are very much more dilute than that.

SAFETY Non-flammable solvents such as dichloromethane are safer in the vicinity of the hot GLC instrument than solvents such as ether both for dissolving the sample and washing out the syringe.

PACKED COLUMNS. The injection is made with a microlitre syringe (usually 10 µl capacity). These syringes are fragile and very expensive; handle with care and do not leave the syringe on the moving chart paper. The length of needle required varies from instrument to instrument but it should be long enough to deposit the sample onto the column packing and not into the empty space above (Fig. 4.3).

To make an injection first make sure that the syringe is clean by dipping the needle into some solvent in a conical flask (not the stock solvent bottle) and carefully filling and discharging it several times. If the plunger action seems sticky or the needle is blocked, see your instructor, as more rigorous cleaning may be needed. Then dip the

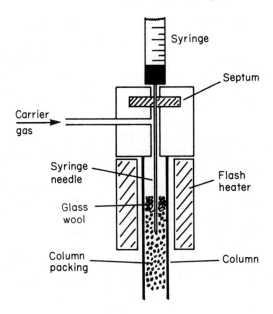

Fig. 4.3 Gas chromatograph injection head.

needle into the sample solution, move the plunger up and down several times to remove air from the barrel and then draw in the amount required (usually 1–5 μl). Wipe the needle on a tissue, check that you do not have any significant air bubbles (it is very difficult and not essential to remove the last trace of air) and make the injection. It is helpful to steady the needle with the left hand while piercing the septum, and it must then be inserted to its full length before injecting. On some instruments there is a marker button which puts a mark on the chart to show the time of injection (e.g. Fig. 4.5). If not, mark the chart with a pencil.

CAPILLARY COLUMNS. These columns can take only *c.* 1% of the load of a packed column. The injection system usually incorporates a stream splitter, which directs a few per cent of the injection onto the column and vents the rest. More modern instruments have splitless injectors (see the instrument manual).

(d) Running the chromatogram

It may be necessary to do one or more trial runs before a satisfactory chromatogram with all peaks on scale is obtained. For the first run you should set the attenuator to a mid-range position (unless your experimental instructions specify otherwise) and then watch the peaks as they appear. If it is obvious that a peak is going off scale then progressively adjust the attenuator as the pen nears the top of the paper until the top of the peak is reached and make a note of the setting used on the chart. Do this for each peak. When the run is complete, look at any small peaks of interest on the chart and work out what attenuation is required to bring them up to a reasonable height (above half scale). Then inject the same volume of sample again and for each peak set the attenuator at the appropriate value before the peak starts to appear*. In some cases all the peaks can be run at the same attenuation, but not often. If the instrument is properly set up (subsection (i) p. 134), changing the attenuation when the pen is on the baseline should not produce any appreciable shift of the pen.

PEAK SHAPE. When running the chromatogram look at the shape of the peaks. Distortion (Fig. 4.4b) from the normal symmetrical shape (Fig. 4.4a) can result from two causes. The simpler to remedy is overloading. This results in a peak for which the pen rises slowly on the forward side and drops steeply back to baseline after the apex. The cure is to reduce the sample size and decrease the attenuation accordingly.

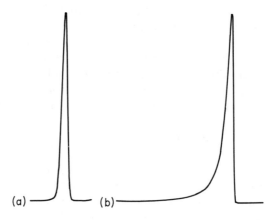

(a)   (b)

Fig. 4.4 Peak shapes in GLC.

---

* Changing the attenuation in this way is not appropriate if an electronic integrator is being used to measure peak areas; see the operating instructions for the integrator.

Less easy to remedy are 'tailing' peaks, which look normal on the forward side but have a long drawn-out 'tail' before the pen finally gets back to baseline. They are usually a symptom of incompatibility between the sample and the stationary phase (see subsection (h) below). However, occasionally they result from an operational fault in sample injection – either due to a cold injection zone (check flash heater setting) or using too short a syringe needle.

The chromatogram of a three-component mixture (all at the same attenuation) is shown in Fig. 4.5. In the next two subsections this will be used to illustrate the use of GLC in qualitative and quantitative analysis.

(e) Identification by GLC

For a given set of conditions (column type, temperature, carrier gas flow rate) every compound has a characteristic retention time (T_1 – T_3, Fig. 4.5).* Therefore, to identify a particular compound in a mixture you must compare its retention time with that of an authentic sample of the compound – run separately. The method of peak enhancement can be used to make certain the retention times are the same. To do this you add a little of the 'authentic' material to a sample of the mixture and run it again. If the peak in question is enhanced in size and remains undistorted then this confirms that the retention times are the same. In a teaching laboratory situation this is generally considered sufficient evidence to identify the compound but in project work and in professional chemistry it would not be regarded as sufficient. The uncertainty arises from the fact that two quite different compounds may by coincidence have the same retention time. If the 'authentic' and the 'unknown' have the same retention time on two quite different stationary phases (subsection (h) below) then this is stronger evidence but even then further points of identity are needed, e.g. a mass spectrum obtained on a coupled gas chromatograph/mass spectrometer. It may be necessary to carry out preparative-scale GLC and obtain an NMR spectrum.

(f) Quantitative analysis

Gas–liquid chromatography is used extensively for the quantitative analysis of mixtures. The procedure is relatively straightforward but

---

* More correctly the retention volume should be calculated, i.e. the volume of carrier gas required to elute the compound (retention volume = retention time × flow rate of carrier gas). However, it is not always possible to measure the flow rate in a teaching laboratory.

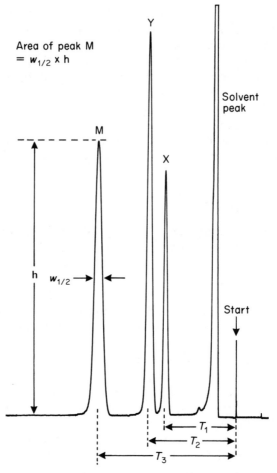

Fig. 4.5 GLC trace of a three-component mixture.

does take some time since it involves calibrating the instrument with pure samples of the compounds to be measured.

The first requirement is to measure the areas of the peaks on the chromatogram. The easiest and best method is to use an electronic integrator but these are not universally available, particularly in teaching laboratories. However, if the peak is reasonably symmetrical, manual measurement will produce good results. An example is shown for peak M in Fig. 4.5. The area is given by multiplying the height (h) by the width at half-height ($w_{1/2}$). To obtain good accuracy in measurement use a fast enough chart speed to avoid producing very narrow peaks. For the width measurement, an illuminated map measurer is easier and much more accurate than a ruler.

The area of a peak is directly proportional to the weight of material passing through the detector. Thus, considering peak X in Fig. 4.5, $W_X \propto A_X$ where W_X is the weight of material producing the peak of area A_X. The proportionality constant relating the two is k_X the response factor of compound X. Thus

$$W_X = k_X A_X$$

In principle k_X could therefore be found by injecting a known quantity W_X and measuring A_X, the area of the peak produced. However, it is very difficult to inject precisely known quantities and the method is not practicable. Two methods are used to circumvent this problem: (*i*) the internal standard method, which is applicable to all mixtures; and (*ii*) the internal normalization method, which is generally used only when all the components have the same response factor.

(i) The internal-standard method

In this a known amount of an internal standard is added to the mixture to be analysed to give a peak separate from the unknown peaks, e.g. peak M in Fig. 4.5. Then the amounts of X and Y present can be worked out by measuring the peak area ratios A_X/A_M and A_Y/A_M. The procedure is as follows:

1. Find a suitable internal standard by trial and error. It must be pure, stable and unreactive to anything in the mixture. Hydrocarbons such as biphenyl, bibenzyl or naphthalene, or esters of various kinds are often used.
2. Add a known amount to the mixture – enough to give a peak of similar size to the unknown peak.
3. Make at least three injections of the mixture until concordant values ($\pm 3\%$) of A_X/A_M and A_Y/A_M are obtained. (If concordant values cannot be obtained, there may be some problem with the instrument (see subsection (i) below) or you may be using solutions that are too concentrated and not on the linear part of the detector response curve.)
4. Considering peaks X and M:

$$W_X = k_X A_X \qquad \text{and} \qquad W_M = k_M A_M$$

and so

$$W_X/W_M = (k_X/k_M) \times (A_X/A_M)$$

The ratio, k_X/k_M is known as the correction factor K_X of compound X. Thus

$$W_X/W_M = K_X(A_X/A_M) \qquad (4.1)$$

In this equation the weight ratio in the sample injection (W_X/W_M) is the same as the weight ratio in the whole mixture irrespective of the size of the injection. Thus if W_M is the weight of internal standard added in step 2 above, then W_X is the weight of compound X in the mixture – the quantity required. Therefore W_X can be calculated from the equation if K_X is known.

The procedure for evaluating K_X is straightforward. A calibration mixture containing known amounts of 'authentic' compounds X and M is prepared and injected and the peak area ratio measured. Then K_X can be calculated from equation (4.1). In practice it is usual to prepare several calibration mixtures and plot a graph of W_X/W_M against A_X/A_M to determine K_X. If several unknown compounds are to be determined then all should be put into the calibration mixture together and a calibration graph plotted for each. The calibration mixtures should be made up to approximately the same dilution as the unknown mixture, i.e. the peaks should be obtained on the same attenuator setting on the chromatograph. At least three injections of each should be made to obtain concordant values of the peak area ratio.

(ii) Internal normalization

This is a less generally applicable method of quantifying GLC analysis, which can only be used if all the compounds in a product mixture are eluted and all give measurable peaks. Unlike the internal standard method this cannot therefore be used when reaction mixtures contain high- molecular-weight tarry or polymeric products, which are not GLC volatile.

The simplest application is where all the components in the mixture have the same response factor. A homologous series of compounds or a mixture of isomers may fall into this category. If so then the component concentrations can be determined from:

$$X(\%) = (A_X/A_T) \times 100$$

where A_X is peak area of component X and A_T is total peak area of all components (excluding solvent peak). It is most important to show that the response factors for the components are the same before using this technique. The method can be modified to take differing response factors into account but in general it is much more satisfactory to use the internal standard method.

(g) Recording GLC data

On each chromatogram the following should be recorded:

1. Experiment number or sample number
2. Injection size
3. Column used, i.e. stationary phase, support material, percentage loading, length
4. Carrier gas flow rate
5. Operating temperature
6. Amplifier attenuation used for each peak
7. Chart speed

An example would be:

Ex. 1, sample 1/4; 1 µl;
7 ft 10% APL on Chromosorb W; 150°C;
N_2, 60 ml min^{-1};
chart, 60 cm h^{-1};
all peaks at 50×10^2.

(h) Selection of the stationary phase and other operating parameters

The procedure for obtaining a good separation for an unknown mixture involves finding the best column and operating temperature by trial and error using the general observations below. As your experience increases it becomes quicker and easier. Overall the procedure follows these lines: Select a column with a particular loading of a particular stationary phase (subsections (i) and (ii)) from the stock available. Set it up in the instrument with the correct gas flow rate (subsection (iv)) and make a guess at an operating temperature (subsection (ii) and (iii)). Make an injection, look at the results and then vary the operating temperature to achieve adequate resolution of the peaks in the minimum analysis time. If the separation of a pair of overlapping peaks cannot be achieved by varying the temperature then it will be necessary to change to another stationary phase and begin again. If there is a known number of compounds in the mixture, it is obvious when the required separation has been obtained. However, in a completely new mixture such as might be obtained in project work, there may be an unknown number of components and it is important to go through the procedure outlined in subsection (ii), to avoid missing any.

(i) Selection of the stationary phase

CONVENTIONAL COLUMN PACKINGS. The general principles involved in GLC separation are discussed in Section 4.3.1(b) on p. 171, which should be read carefully before this section is used. Basically the general expectation is that more volatile materials will pass through the column faster than less volatile ones. The compounds injected will therefore be eluted in order of increasing boiling point. However, this order can be perturbed considerably by specific interactions between the compounds and the stationary liquid phase. These interactions depend on the nature of the compound (polar/non-polar) and the nature of the stationary phase, whose properties can also be very loosely defined in terms of 'polarity'. The general types are listed in Table 4.1. Several hundred stationary phases are commercially available but many have similar properties. Most laboratories keep only a few examples of each type and this suffices for the majority of separations. Some specific examples listed in order of increasing polarity are given in Table 4.2.

Each particular liquid phase has a maximum recommended temperature of operation (Table 4.2) above which excessive column bleed (loss of stationary phase) or thermal degradation will occur.

The choice of the correct stationary phase for a particular separation is not easy and can involve much trial and error. There are many literature compilations of retention data which can be useful when dealing with known compounds or mixtures, but for new analyses the chromatographer has to rely on a few basic guidelines and on the

Table 4.1 General types of stationary phase for GLC

Non-polar	Apiezon greases, e.g. APL silicone gum rubbers, e.g. SE30/E301, OV1 (dimethyl) silicone fluids, e.g. OV17 and OV25 (phenylmethyl)
Intermediate polarity	silicones with polar groups, e.g. XE60 (cyanoethylmethyl), QF1 (fluoro) polyphenyl ethers
Polar	polyethyleneglycols (Carbowaxes) in a range of molecular weights polyamides, e.g. Versamide 900 esters and polyesters, e.g. neopentylglycol succinate (NPGS), diethyleneglycol succinate (DEGS), polyethyleneglycol adipate (PEGA)

Table 4.2 Some common stationary phases for GLC

Stationary phase	Max. temp. (°C)
squalane	150
Apiezon L	250
silicone gum (methyl) SE30	250
silicone (methyl) OV1	300
silicone gum (phenyl) SE52	250
silicone (phenyl) OV17	300
silicone (phenyl) OV25	300
silicone (fluoro) QF1	225
silicone gum (nitrile) XE60	225
poly-*m*-phenyl ether (6 ring)	250
neopentylglycol succinate (NPGS)	200
Carbowax 20M	220
Carbowax 1000	150
ethyleneglycol phthalate	200
diethyleneglycol succinate (DEGS)	190
cyanoethylsucrose	125

Rohrschneider McReynolds numbers[3a,b], which give a quantitative measure of polarity and an indication of selectivity. Generally 'like retains like', e.g. on 'non-polar' columns non-polar compounds are eluted more slowly than polar compounds of the same boiling point. On 'polar' columns polar compounds are selectively retained over non-polar ones of the same boiling point.

Polar compounds such as alcohols, amines and carboxylic acids can be difficult to analyse by GLC as they tend to give 'tailing' peaks (Fig. 4.4b). This results from strong adsorption of the compound onto active sites on exposed parts of the surface of the support material. This problem can be avoided at least in part by using a column packing in which the support material has been deactivated by acid washing and silanization (such supports are often marked AW/DMCS). This is particularly important for lightly loaded columns containing less than 5% of the stationary phase. The non-conventional packings discussed below are particularly suitable for polar compounds.

3a. Supina, W. R. and Rose, L. P. (1970), *J. Chromatogr. Sci.*, 8, 214.
3b. McReynolds, W. O. (1970) *J. Chromatogr. Sci.*, 8, 685.

POROUS POLYMER BEADS. These materials serve as both support and stationary phase and are available in various types (Waters 'Porapak' range and Chromosorb 'Century' packings). They are excellent for the separation of polar compounds with minimum tailing and for gas analysis.

BONDED STATIONARY PHASES. Stationary phases can be permanently bonded to the surface of the support material. In some cases the loading is very low (0.2%) (see below) and the surface is so completely deactivated that polar compounds do not give tailing peaks. Bonding, however, can change the separation characteristics of the stationary phase.

(ii)   The percentage loading of the stationary phase and the operating temperature

The amount of stationary phase coated onto the support material usually varies between 1 and 10% by weight for conventional analytical columns. The loading can go as high as 25% but above that the packing becomes sticky. At a given temperature the retention volume of a compound is proportional to the amount of stationary phase so, for example, a compound will pass much more quickly through a column with a 1 or 2% loading than one with 10%. Lightly loaded columns (1–5%) are therefore of great value in the analysis of high-molecular-weight compounds of low volatility, which would require an inconveniently long analysis time on say a 10% column.

The rate at which compounds pass through the column is also affected by the column oven temperature. This is the most easily controllable operating parameter on the GLC instrument. The higher the temperature the shorter the retention time for a given compound. However, raising the temperature also brings adjacent peaks closer together so, in practice, temperature optimization is a compromise between shortening analysis time and maintaining adequate resolution between the components of interest.

The optimum temperature and percentage loading of stationary phase for a particular separation can only be arrived at by trial and error but the following will serve as a guide:

1. A column with a 10% loading of stationary phase should be operated *c.*. 30–50°C below the b.p. (760 mmHg) of the compounds being separated. Columns with lower loadings of stationary phase will of course require lower operating temperatures to give similar retention volumes.
2. Thermally unstable compounds are best separated with lightly

loaded columns so as to keep the operating temperature as low as possible.

3. High-molecular-weight compounds of low volatility require lightly loaded columns so as to keep the operating temperature below the maximum allowed for the liquid phase, e.g. tetraphenylnaphthalene (MW 432) can only be chromatographed on a very lightly loaded (1%) column at *c.* 220°C.

4. Conversely, very volatile compounds may need a heavily loaded column (10–25%) to give acceptable retention times at the minimum operating temperature of the chromatograph (usually room temperature).

5. Temperature programming can be used. For mixtures containing components with a wide range of boiling points the column oven can be programmed to start at a low temperature, so as to achieve separation of the volatile components, and to increase the temperature as the analysis proceeds so that the less volatile components are eluted in a reasonable time (Fig. 4.6). Temperature programming is very valuable for examining mixtures of unknown products from new reactions when it is most important that the column temperature is programmed over the whole available range up to the column's temperature limit. The mixture should be examined with both high and low percentage stationary phase columns. Otherwise it is very easy to miss either very volatile products owing to their elution with the solvent peak or very involatile components with long retention times, which may be left on the column.

(iii) Injection temperature and detector temperature

In most gas chromatographs the 'injection' end of the column can be locally heated above the oven temperature so as to provide 'flash' vaporization of the sample (Fig. 4.2a). In general the flash heater should be set at *c.* 50°C above the oven temperature but care must be taken to avoid the thermolysis of labile compounds. For the latter the injection heater should be set at the column oven temperature and the sample should be injected directly onto the column packing with a long-needle syringe. Some instruments have the detector in a separate oven; this should also be set at a temperature about 50°C above that of the column oven.

(iv) Carrier gas flow rate

The type of gas does not have a major effect on resolution but the flow rate is very important as the linear gas velocity affects column efficiency. Efficiency is low at low flow rates but increases with flow

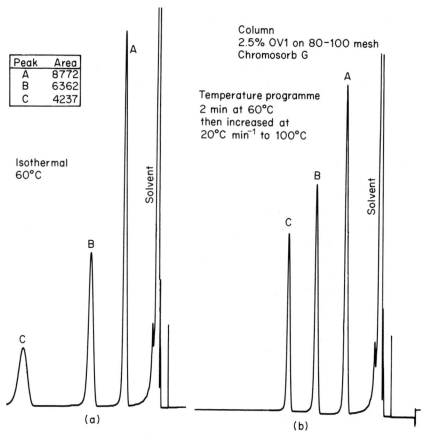

Peak	Area
A	8772
B	6362
C	4237

Column
2.5% OV1 on 80–100 mesh
Chromosorb G

Temperature programme
2 min at 60°C
then increased at
20°C min⁻¹ to 100°C

Isothermal
60°C

Fig. 4.6 GLC traces (a) without and (b) with temperature programming.

rate to a broad maximum and then decreases again at high flow rates. Columns of 4 mm bore are usually operated with a flow of c. 40–60 ml min⁻¹, 2 mm bore columns with c. 15–20 ml min⁻¹, and capillary columns 1–2 ml min⁻¹.

(i) Setting up the instrument – some general points

Instruments differ and each should be set up in accordance with the instruction manual. There are, however, a few general points to be checked to ensure good performance. In quantitative work in particular it is vital that the correct gas flow rates are used both for the column and the flame ionization detector (see manual). The response of the detector can be variable, inconsistent and non-linear if the air supply is inadequate and the flame will not be stable unless the hydrogen flow is correct. The system must be completely leak-free; even small leaks around the septum or at the column connections will lead to irreproducible results, so the connections should always be leak-tested with a detergent solution (*c.* 10% aqueous liquid detergent) before the oven heater is turned on. It is also essential that the electrical system is properly set up. Each instrument will have its own procedure but it essentially consists of aligning the zero levels of the amplifier and recorder and then backing off the background standing current from the detector. When properly set up the recorder pen should not move significantly from the baseline when the attenuation is changed.

After being in use for some time GLC columns may become contaminated with small amounts of very involatile compounds that are eluted very slowly in normal use and produce an unstable, undulating baseline on the chart recorder. Such contaminants can be removed by 'purging' the column. This involves heating it, usually overnight, at its maximum operating temperature (Table 4.2) with a slow flow of carrier gas but with the exit end disconnected from the detector.

4.1.3 High-performance liquid chromatography

High-performance liquid chromatography (HPLC) is a small-scale 'instrumental' form of column chromatography. It is an extraordinarily versatile technique, which can be used for the qualitative and quantitative analysis of mixtures of all kinds of compounds – volatile or non-volatile, thermally stable or not, polar and non-polar.

(a) General description

The essential features of the equipment used are shown in Fig. 4.7. The column contains the stationary phase – usually an adsorbent (e.g. silica or alumina) or a bonded liquid phase (see p. 174) – in the form of very fine particles*. The liquid mobile phase (eluting solvent) is supplied by a high-pressure pump. A solution of the mixture is injected onto the top of the column packing with a microlitre syringe. The compounds in the mixture are then carried down the column by the eluting solvent with different compounds moving at different rates depending on their relative affinity for the stationary phase and the moving eluent. Ideally they separate completely into discrete

---

* Other modes of chromatography – ion-pair, ion-exchange or size exclusion – are also used but are not covered here.

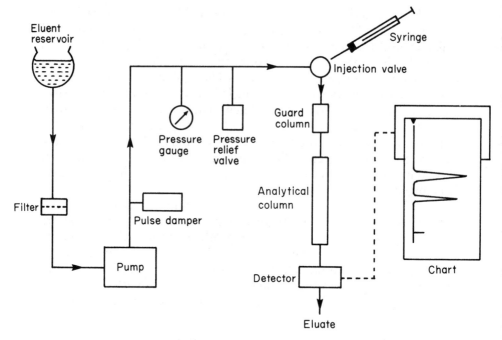

Fig. 4.7 Apparatus for HPLC.

bands (as shown for a preparative scale column in Fig. 4.16) which are eluted in sequence from the bottom of the column into a detector. More details of the separation processes are given in Section 4.3.1. The detector responds to each compound passing through it and produces a peak on a chart recorder or on very modern instruments on a visual display unit (VDU). The output is therefore of essentially the same form as in GLC. A typical separation is shown in Fig. 4.13. The technique outlined in this section applies to analytical separations in which submilligram quantities of sample are used per injection. These separations rarely take more than 30 min and often very much less.

(b) Equipment

A typical HPLC set-up (Fig. 4.7) consists of the following pieces of equipment.

(i) The column

The HPLC column is made from stainless-steel (SS) tubing (0.2–0.5 cm bore × 8 –25 cm long), which is packed with the stationary phase retained between two fine SS gauze discs. There are many different types of end fitting for making the connections to the solvent pump and the detector. Once packed these columns have a long life time if treated carefully. The column is usually connected to a short guard column at the inlet end, which protects the expensive analytical column from dust, tars and other impurities that may be present in the sample. The guard column is packed with the same material as the main column.

(ii) The solvent delivery system

The solvent is pumped through the column usually by a piston-type pump capable of pressures up to 3000–6000 psi, but for many applications pressures of only a few hundred pounds per square inch are required. Pumps of this type require a pulse damper in the system to even out the solvent flow between the piston strokes.

The simplest arrangement is a single-pump system (Fig. 4.7), used for isocratic (constant-solvent-strength) elution. More advanced systems are capable of gradient elution – the controlled variation of solvent strength during the separation. In this case two pumps are used (Fig. 4.8), one supplying the strong and the other the weak solvent, first to a mixing chamber and thence to the column. The ratio of the two solvents is controlled by a microprocessor or a small computer, which can be programmed to adjust the ratio either in steps or as a smooth gradient as appropriate to the particular separation.

In all cases the solvents must be degassed before use to minimize

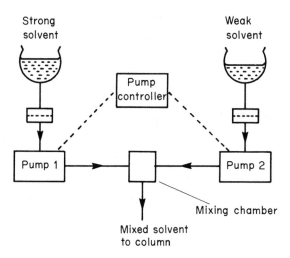

Fig. 4.8 Two-pump system for gradient elution in HPLC.

the formation of air-bubbles in the detector cell and filtered to remove dust (a filter is usually incorporated into the feed line to the pump).

(iii) The detector

Detectors monitor some property of the eluate from the column and produce an electrical signal when a compound emerges in the solvent stream. Two types are in general use – ultraviolet (UV) and refractive index (RI) detectors. The former is by far the most common and, operating at 254 nm, will detect almost all types of compound except those that are fully saturated. Obviously the eluting solvent must be transparent to UV. Refractive index detectors are less sensitive but are invaluable for simple and complex saturated compounds. They work in a differential mode, comparing the refractive index of the eluate from the column with that of the pure solvent.

(iv) The injection system

The sample mixture is loaded onto the column as a solution which can be injected either directly onto the top of the column (Fig. 4.9) or introduced into the eluent stream via an injection valve (Fig. 4.10).

SEPTUM INJECTION. The method can only be used at solvent pressures up to *c.* 1500 psi. The column is fitted with an injection

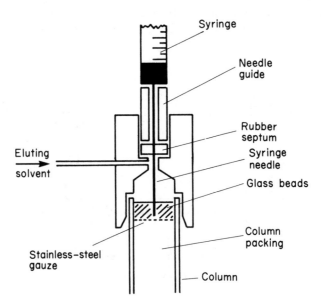

Fig. 4.9 HPLC injection head.

head (Fig. 4.9) containing a PTFE-faced silicone rubber septum. The correct length of needle must be used so that the sample is injected right at the centre just above the top gauze – usually into a bed of fine glass beads above the column packing. The beads are renewed periodically to remove bits of the septum that break off in use. The method provides high resolution but suffers some disadvantages, notably that the syringe needle frequently gets blocked with the septum rubber. This can be minimized by using pre-drilled septa and a blunt dome-tipped needle but can still cause problems particularly with inexperienced operators.

VALVE INJECTION. Injection valves vary in design but the general principle is shown in Fig. 4.10. In the 'load' position the sample solution is loaded at atmospheric pressure into a sample loop

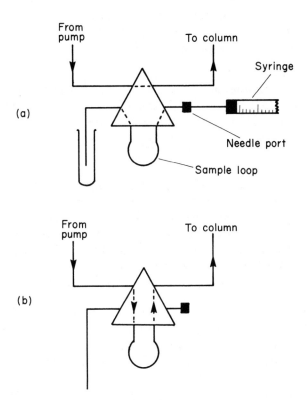

Fig. 4.10 Injection valve for HPLC (schematic): (a) in 'load' position; (b) in 'inject' position.

while the eluent is being pumped directly to the column. When the valve is moved to the 'inject' position the eluent is diverted through the loop to wash the sample onto the column. For analytical work sample loops from 5 to 20 μl are suitable but larger ones are available so that the valve can be used for preparative work with bigger columns. The valve can be used in two ways: the sample loop can either be filled completely with the sample solution or, alternatively, a small volume (a few microlitres) can be injected into a large loop (20 μl) using a microlitre syringe and a special needle seal attached to one of the ports. The latter method is convenient for exploratory work since it is easy to vary the sample size, but the former provides a wholly reliable method for achieving reproducible injections.

(c) Operation of the equipment

(i) Setting up

The operation of HPLC equipment is very straightforward once it has been set up and optimized for a particular separation. In its simplest form, using a single pump (Fig. 4.7), the sequence is as follows:

SAFETY The solvents used are usually volatile and flammable, so take appropriate precautions (Section 1.3) to prevent ignition and inhalation. All mixing operations should be done in a fume-cupboard.

---

1. De-gas the eluting solvent (or solvent mixture) before putting it into the solvent reservoir. The simplest method is to apply a vacuum (water pump) to the solvent in a flask while stirring (magnetic) or shaking the contents. This is a quick method but tends to disturb the balance of mixed solvents due to preferential evaporation of the more volatile one. Alternatively, and preferably, the solvent can be boiled under reflux, allowed to cool undisturbed and used at once.
2. Set the solvent flow rate to the prescribed value (this depends on the column diameter, e.g. $1-3$ ml min^{-1} is appropriate for a 5 mm bore column), and switch on the pump.
3. Allow the solvent to flow through the column and detector until no more air bubbles are seen in the eluate.
4. During this period switch on the detector and set the range switch to its least sensitive position.
5. Switch on the chart recorder.

After a few minutes the baseline should have stabilized. When it has, check it over the full usable span of the range switch. If there is any problem at this stage, consult your instructor, as it may be due to column contamination or more simply the entrapment of an air bubble in the detector flow cell.

(ii) Running the chromatogram

The concentration of the sample solution should be adjusted so that the amount of material injected falls in the range from a few micrograms up to *c.* 0.5 mg. The sample should be dissolved in a solvent that will not give a peak (or one of minimum size) on the chromatogram – a UV-transparent solvent if UV detection is being used, or for a refractive index detector, use a little of the eluting solvent where possible. The best resolution is obtained with small sample sizes and low-volume injections. For syringe injection – either direct or into the sample loop of a valve injector – the volume is usually in the 1–5 µl range, and for valve injection, using the full loop capacity, 5 or 10 µl.

SEPTUM INJECTION. Washing out and filling syringes is described in the GLC section (Section 4.1.2). In septum injection the needle has to be inserted through a septum (Fig. 4.9), which is firmly compressed, and then the solution has to be injected against a very high back-pressure (up to 1500 psi). Insertion of the needle is made easier if the septum has been pre-pierced with a larger needle during installation or if a pre-drilled septum is used. When making the injection it is essential to hold the plunger firmly to prevent it being blown back out of the barrel by the solvent pressure. Insert the needle fully before injecting. Valve injection is much easier.

VALVE INJECTION. The general principles of valve operation (Fig. 4.10) are outlined in subsection (b) on p. 139, but details vary depending on the particular make and type. First wash out the sample loop using a large syringe. Check that it is clean by injecting a loop full of the pure solvent. If no peaks appear other than the small one due to the solvent (which may be positive or negative), then proceed with the sample injection.

OPTIMIZATION OF SAMPLE SIZE AND RANGE SETTING. Make the first injection at a mid-range setting of the range (attenuator) control on the detector. As the peaks appear, adjust the setting to bring each peak on scale on the recorder and make a note of the setting beside the peak. Non-linear effects are sometimes a problem with high sample concentration, so if any of the peaks require the minimum (least sensitive) setting, then either dilute the sample or use a smaller injection size for the next run. Also, look at the shape of the peaks: an unsymmetrical shape can indicate overloading of the column (or detector) and is another indication to reduce the sample size (or concentration). Remember also that resolution gets better as the amount of sample is reduced. In the limit, however, baseline noise will be a problem at the very sensitive range settings needed for very low sample concentrations. It may take several runs to get the sample size and range settings optimized. For the final run set the range setting to the appropriate value before each peak appears*.

(d) Qualitative and quantitative analysis

(i) Identification of compounds

As in GLC a compound has a characteristic retention time (or retention volume) (Fig. 4.11) under a particular set of operating conditions. This can be used for provisional identification by comparison with an 'authentic' sample. The methods described in the GLC section (Section 4.1.2) apply equally to HPLC and the results are subject to the same degree of uncertainty. However, in HPLC it is easier to obtain supporting evidence by collecting the sample for mass spectrometry.

(ii) Quantitative analysis

The internal standard method (Section 4.1.2) is equally applicable to HPLC. When using a UV detector it should be set at a wavelength where all compounds concerned show a reasonably strong absorption.

It is important to realize that rough estimates of the relative amounts of compounds in a mixture based simply on relative peak areas can be wildly wrong particularly when using UV detection. This arises because the compounds may have enormous differences in their extinction coefficients at the detector wavelength being used.

(e) Retention and resolution – basics

This section contains definitions of some of the terms used in HPLC and provides an indication of how resolution is affected by the variables that are under the control of the operator. In the following it is assumed that the reader has looked at Section 4.3.1, p. 170, which outlines the general principles of separation.

---

* See footnote on p. 124

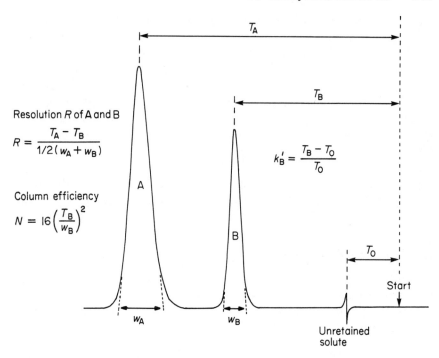

Fig. 4.11 HPLC retention parameters.

The separation of two components, A and B, is shown in Fig. 4.11. Each peak has a characteristic retention time (T) under a given set of conditions, but this is generally converted into a retention volume V where

$$V = T \times \text{(eluent flow rate)}$$

The 'unretained' peak is due to non-polar material (sample solvent) which has passed through the column at the same rate as the eluting solvent. Its retention volume V_0 is known as the void volume of the column – sometimes simply referred to as the column volume. Separations in HPLC are usually discussed in terms of the k' value of each compound (see Section 4.3.1(a)). The k' values can easily be obtained from the retention data, e.g.

$$k'_A = (V_A - V_0)/V_0 \qquad \text{or} \qquad k'_A = (T_A - T_0)/T_0$$

Thus k'_A provides a measure of the retention of compound A (in column volumes) and is known as the capacity factor of A. Obviously the bigger the difference between k'_A and k'_B (Fig. 4.11) the better the separation between them. A measure of this is given by their ratio

k'_A/k'_B, which is known as the separation factor α. For good separations α should be as large as possible.

The resolution R of the two peaks – the measure of how well they are separated – depends not only on $T_A - T_B$ (Fig. 4.11) but also on the width of the peaks. The latter is related to the column efficiency, which is usually expressed as the number of theoretical plates N. The resolution can be expressed in terms of the three operator variables, α, N and k':

$$R = \tfrac{1}{4}(\alpha - 1)N^{1/2} [k'/(1 + k')] \qquad (4.2)$$

where k' is the average of k'_A and k'_B, α $= k'_A/k'_B$ and N is the number of theoretical plates of the column. Equation (4.2) provides the insight into how a separation can be optimized by the operator.

(f) Optimizing the separation

Consider a situation where the initial attempt to separate A and B has not produced complete resolution (Fig. 4.12). We examine below what the operator can do to improve the situation by manipulating N, k' and α.

(i) Number of theoretical plates N
The efficiency of the column depends essentially on the size of the packing particles and how well it has been packed. However, for

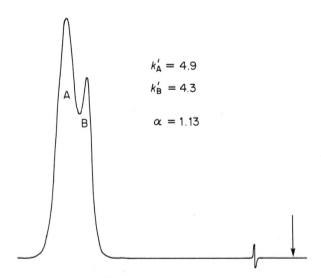

$$k'_A = 4.9$$
$$k'_B = 4.3$$
$$\alpha = 1.13$$

Fig. 4.12 Incomplete resolution of two components in HPLC.

a given column it does also depend on flow rate, which is an easy parameter for the operator to adjust. The best value depends on the column internal diameter (i.d.): $0.1-1$ ml min^{-1} for $2-3$ mm, and $1-3$ ml min^{-1} for $3-5$ mm columns. However R depends only on $N^{1/2}$ so the improvement may not be spectacular.

(ii) Capacity factor k'

The retention volumes of the compounds (and hence the k' values) depend on the strength of the eluting solvent (see Table 4.5, p. 172). By weakening the solvent the retention volumes can be increased. The key question is whether this also improves the resolution. The answer is yes, but with limitations. The resolution R is proportional to $[k'/(1+k')]$; therefore as k' is increased R goes up rapidly at first but then more slowly and increases little more after $k' = 5$. Thus there is no point in excessively weakening the solvent ($\varepsilon°$) and extending the retention time since there will be little further gain in resolution. The optimum k' range is between 1 and 10 when the samples will be eluted by $2-11$ column volumes of solvent.

(iii) Separation factor α

Consider the situation where the solvent strength has been adjusted to bring k' into the optimum range but the two compounds are still not resolved. By definition $\alpha = k'_A/k'_B$. The individual k' values for the compounds depend on their distribution ratios between the stationary and mobile phases (Section 4.3.1a), and these (and hence α) can be changed by changing the constituents in the mobile phase, owing to selective solute–solvent interactions. Therefore the technique is to hold the polarity (strength) of the solvent more or less constant – preserving the optimum k' – but to vary the constituents in the mobile phase in the hope of changing α. This is discussed further in later sections on method development for the various modes of chromatography.

(g) Selection of the chromatographic mode

The HPLC methodology can be used for various types of chromatography*. The common ones are liquid–solid (adsorption) chromatography (LSC) and liquid–liquid chromatography (LLC) using bonded liquid stationary phases. Adsorption chromatography (LSC) is applicable to a wide range of non-ionic compounds from moderately non-

---

* Ion-exchange, ion-pair and size exclusion chromatography are not covered in this treatment.

polar to moderately polar but does not work very well for either very non-polar species (hydrocarbons) or very polar ones such as amines, alcohols or acids. Bonded-phase LLC in its normal and reverse-phase modes can cope with almost all compound types and is especially suitable for polar compounds, which cannot be handled by LSC. This and the fact that LLC is in several ways operationally easier than LSC accounts for its enormous growth in popularity over the last five years.

(h) Liquid–solid (adsorption) chromatography (LSC)

The general principles of separation are outlined in Section 4.3.1(a).

(i) Adsorbents

The adsorbent used is almost invariably silica or alumina specially prepared in the form of porous spherical particles in sizes from 3 to 10 μm. The use of these very small particles is the key to the high resolution obtained in HPLC. A typical 'wide-bore' column 10 cm long × 4 mm bore would give c. 6000 theoretical plates* when packed with 5 μm material and c. 10 000 plates with 3 μm material. Columns are usually bought ready-packed to a specified performance.

Both silica and alumina can be used for separating a wide range of compound types. Very strong compound – adsorbent interactions should be avoided since they can give rise to bad peak tailing. Thus amines are better analysed on alumina than on silica (which is acidic) and phenols better on silica than on alumina. Very polar compounds do not give good results and are better analysed by bonded-phase LLC (following subsection (i) on p. 150).

A preliminary examination by TLC may indicate which adsorbent is best for a particular sample but in cases where resolution cannot be achieved by TLC both types of HPLC column may have to be tried.

(ii) Control of adsorbent activity

The surface of the adsorbent is covered with sites that have a wide range of different adsorption strengths. To obtain good peak shapes and reproducible retention times it is essential that the most highly active sites are blocked off with some strongly adsorbed compound such as water or an alcohol. The activity of the surface depends on what proportion of the adsorption sites are blocked. Water deactiva-

---

* The efficiency of a chromatography column, like that of a distillation column (Section 3.4.3) is measured in theoretical plates – the higher the number, the better the resolution (see Fig. 4.11).

tion is commonly used since it is wholly reliable when carried out properly. However, it is time-consuming since it requires careful control of the water content of the eluting solvent. Other polar surface 'modifiers' can be used instead (see below).

WATER DEACTIVATION. Coverage of about half of the surface by a water monolayer is usual. The column can be 'conditioned' to this level of activation by passing through it about 10–20 column volumes of ether containing the required amount of water. Silica requires *c.* 50% water-saturated ether and alumina requires 25% water-saturated ether[4]. These are prepared by mixing dry ether with fully water-saturated ether in the appropriate proportions. As the wet ether passes down the column it equilibrates with the adsorbent either taking up or releasing water until equilibrium is reached[*]. The percentage water saturation can be varied to produce the activity required, e.g. the separation of the four isomeric hydrocarbons (Fig. 4.13) was achieved after conditioning the alumina with 15% water-saturated ether, whereas no resolution was obtained on the less active column obtained by conditioning with the more usual 25% water-saturated ether. It is thus fairly quick and easy to condition the column to the required activity. The problem lies in keeping the activity constant throughout the whole period of an analysis. All eluting solvents can take up water from the adsorbent or release it depending on their dryness. To keep the activity constant the eluent must be prepared to the same water activity as the original conditioning solvent. This involves preparing the eluent in both dry and fully water-saturated states and mixing them in the appropriate proportions.

OTHER SURFACE MODIFIERS. Organic modifiers such as methanol, isopropanol or acetonitrile can be used as surface modifiers instead of water. They are added to the eluting solvent in tiny amounts (0.01–0.5%) to set up and maintain the required activity. The amount needed varies with the nature of the modifier and the eluent, e.g. the addition of 0.15% methanol or 0.3% isopropanol to dichloromethane produces the same order of activity as 50% water saturation. It has been reported that in some cases the use of modifiers other than water produces poor peak shapes.

4. Snyder, L. R. and Kirkland, J. J. (1979) *An Introduction to Modern Liquid Chromatography*, 2nd edn, Wiley-Interscience, New York.

[*] This 'conditioning' of the column with ether also serves to flush off any contaminants left on the column from earlier analyses. It is good practice to continue until a stable baseline is obtained.

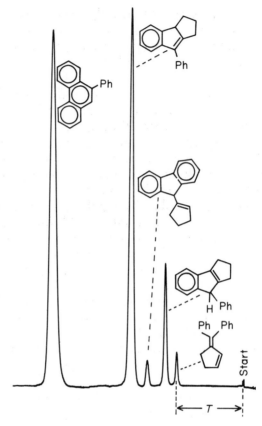

Fig. 4.13 HPLC separation of a mixture of hydrocarbons on alumina.

(iii) The mobile phase

Eluents are selected from more or less the same range of organic solvents used for TLC and preparative column chromatography (Table 4.5, p. 172). Obviously UV-absorbing solvents are not suitable when a UV detector is used. Specially purified and dried HPLC-grade solvents are available. It is rare that a single solvent will have the optimum strength (see Section 4.3.1(a)), for a particular analysis and it is usual to use mixtures of hexane as the weak solvent with an appropriate strong solvent such as ether to give the required $\varepsilon°$ value (Fig. 4.23). For each new analysis the choice of the best solvent strength and of which solvents to use to produce it are of key importance and provide the most challenging operational aspect of HPLC. The line of approach to this problem is discussed in the next section.

(iv) Strength and composition of the eluting solvent

The objective of adjusting the composition of the mobile phase is to achieve or improve resolution between the components and to minimize the analysis time.

The initial solvent strength to be tried can be estimated by a preliminary TLC examination. The objective is to find a solvent mixture (usually hexane and a strong solvent; see Table 4.5, p. 172) that will produce an R_f of 0.2–0.3 for the compounds to be separated. If the mixture contains compounds of widely different polarity then stepped or gradient elution will be required – see last paragraph of this section.

It is generally better to start with a solvent mixture that is strong enough to elute all the components fairly rapidly and then to weaken it (by increasing the percentage of the weak solvent) to optimize the separation following the guidelines in Section 4.1.3(f). If k' for a pair of components has been optimized but separation has still not been achieved, the technique is to hold the polarity of the solvent (ε°) more or less constant – preserving the optimum k' – but to vary the constituents in the mobile phase in the hope of changing α. For example, two components might have an α value of 1.05 using a mobile phase of 50 vol% ether in hexane and so might be incompletely resolved, while the use of 6 vol% dioxane in hexane could give an α of 2 and complete separation, yet both solvent mixtures have the same polarity ($\varepsilon^\circ = 0.3$).

The variation of α using Table 4.5 and Fig. 4.23 as a guide is where the trial-and-error aspect comes in, for there is no wholly reliable way to predict how α will change with eluent composition. However, Snyder[5] has shown that solvent mixtures that contain either very small or very large concentrations of the strong solvent are more likely to give high α values than intermediate solvent concentrations. For example, in the separation of 1-acetylnaphthalene and 1,5-dinitronaphthalene, a solvent mixture of 5 vol% pyridine in pentane gave an α of 2.4 while 25 vol% ether in pentane gave an α of 1.16 although both mixtures have an ε° of 0.25.

GRADIENT ELUTION. In some cases mixtures contain both fast-running low-polarity compounds and more strongly retained polar species. The analysis then requires the use of either step or gradient elution. Good gradient systems (Fig. 4.8) can be programmed to increase the concentration of the stronger solvent in steps, gradients or

5. Snyder, L. R. (1971) *J. Chromatogr.*, **63**, 15.

combinations of both. However, even with simple pumps the solvent composition can be manually changed in steps to allow the elution of polar components. Solvent purity is more critical in gradient elution; for example when using UV detection any UV-absorbing contaminant in the stronger solvent will cause baseline drift as its concentration increases. Any polar impurities in the weak solvent may be absorbed on the column at low $\varepsilon°$ values but later eluted as peaks or humps in the baseline when the solvent polarity has increased.

(i) Liquid–liquid chromatography (LLC) on bonded phases

The general principles of LLC are outlined in Section 4.3.1(c). It is a powerful technique, which can be used across a wide range of compound types, including those not easily separated by adsorption chromatography (LSC), e.g. hydrocarbons (which are not well retained on adsorbents) and polar compounds (which do not give good peak shapes because of strong adsorption). Another major advantage of LLC is that there is no need to 'condition' the column and control its activity as in LSC (see previous subsection (h)).

(i) The stationary phase

The stationary phase consists of a monolayer of an organic liquid chemically bonded to the surface of a silica particle as shown in Fig. 4.25, p. 175. Particle sizes are similar to those used in LSC (3–10 μm). The bonded phases come in two types: simple long-chain hydrocarbons (e.g. C_8H_{17} or $C_{18}H_{37}$), which provide a highly lipophilic layer at the particle surface, and analogues in which the chain is shorter and contains a polar group such as cyano, amino or diol. There are two distinct modes of operation: 'reverse-phase' with the former and 'normal-phase' with the latter.

(ii) Reverse-phase LLC

Here the stationary phase is non-polar – the lipophilic hydrocarbon monolayer on the particle surface – and the eluent is highly polar – usually a mixture of water with a polar organic solvent such as methanol, acetonitrile, dioxane or tetrahydrofuran. This is the reverse of the situation in adsorption chromatography – hence the name. When polar compounds are injected onto the column they have a greater affinity for the polar eluent than for the non-polar stationary phase and therefore pass through rapidly whereas non-polar compounds partition preferentially into the stationary phase and are retained longer. Thus the order of elution is reversed in comparison with LSC, polar compounds being eluted early and non-polar ones late.

OPERATION AND METHOD DEVELOPMENT. The first objective is to find an organic : water ratio for the eluent that will give a reasonable retention time for the compounds to be separated. This involves trial and error and can be done much more quickly if a two-pump system (Fig. 4.8) is available. Useful information on the most suitable organic component and the percentage composition of the eluent for compounds of various types can be obtained from the applications literature (for example, see reference 6). However, in the absence of other indications, it is usual to use methanol as the organic component for the exploratory work. Remember that this is reverse-phase: polar compounds are eluted fastest by weak eluents in which the water content is high but non-polar compounds require stronger eluents that are rich in the organic component.

If peaks are unresolved then the solvent strength can be reduced to extend the retention times (and the k' values) but as in LSC there is little point in going much beyond a k' value of 10 (see previous subsection (f) on p. 144). If peaks are still unresolved the organic component should be changed in an attempt to vary the separation factor α – keeping ε° more or less constant.

One unexpected variable in bonded-phase LLC is the column packing. Samples of what are nominally the same stationary phase from different manufacturers (and even different batches from the same manufacturer) can show different separation characteristics. This is due to the presence of residual silanol groups on the silica surface.

The basic reverse-phase technique described here can be modified in a variety of ways to extend its range by the addition of buffers, acids, bases or detergents to the mobile phase (see reference 6 below, and reference 2 on p. 171).

(iii) Normal-phase LLC

In this form of LLC the bonded stationary phase has polar groups[*] and the eluent is hydrocarbon-based as in adsorption chromatography (LSC). These columns can be used as an alternative to silica LSC, with the attraction that the bonded phases give a wide range of selec-

---

6. Pryde, A. and Gilbert, M. T. (1979) *Application of High Performance Liquid Chromatography*, Chapman and Hall, London.

[*] Some polar bonded phases, e.g. cyanopropyl, can be used in both normal- and reverse-phase modes.

tivity. They are particularly suitable for polar compounds and give better separation and peak shape than LSC. The eluent is usually based on hexane or 2,2,4-trimethylpentane, with a strong solvent (Table 4.5) added to produce the required $\varepsilon°$ value as in LSC. Optimum solvent strength and solvent composition again have to be found by trial and error and the comments on method development on p. 149 also apply.

4.2 PREPARATIVE METHODS

4.2.1 Preparative thin-layer chromatography

This is a version of analytical TLC (Section 4.1.1) scaled up to allow separations on a 'preparative' scale. The plates used (e.g. Fig. 4.14) are larger than analytical plates, typically 20 × 20 cm, and carry a thicker layer of adsorbent (usually 0.5 or 1 mm). A plate of that size could be used to separate up to c. 100–150 mg of a mixture, depending on the difficulty of the separation.

The sample solution is applied as a band along the origin. It can be put on as a series of overlapping spots using a dropper or syringe or as a continuous band by using a cut-down Pasteur pipette with a cottonwool wick (Fig. 4.15). Both methods require the use of a straight-edged guide raised just above the adsorbent surface. Great care is needed to produce a narrow uniform band. A mechanical 'streaker' is preferred if there is much preparative TLC work to be done. This device has a carriage which transports a syringe across the plate, automatically dispensing the solution as it goes along. To keep the band narrow it is best to build up the sample loading in stages by

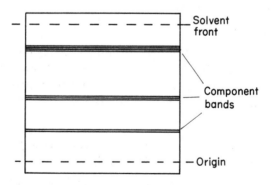

Fig. 4.14 Preparative TLC.

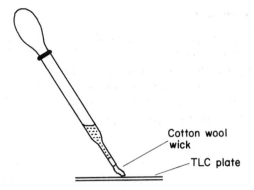

Cotton wool wick

TLC plate

Fig. 4.15 Loading a preparative TLC plate.

applying several light 'streaks' one on top of the other, allowing each one to dry before the next is applied.

After development (Section 4.1.1) the component bands are usually located by the UV method (Section 4.1.1). Spray reagents and iodine would destroy the sample. The band positions are marked and then the adsorbent is carefully scraped off the plate.

SAFETY This operation must be carried out in a fume-cupboard

The compounds are then extracted by putting the scraped-off adsorbent into a sintered glass funnel and adding an appropriate solvent, usually dichloromethane or ethyl acetate depending on the polarity of the compound.

4.2.2 Column chromatography

Column chromatography is the organic chemist's single most important technique for the separation of mixtures on a preparative scale (a few milligrams up to tens of grams). The separation process used is usually liquid – solid (adsorption) chromatography, since it works well for most non-ionic types of compound. The general principles of separation are outlined in Section 4.3.1. The same basic experimental methodology is also used in ion-exchange and size exclusion chromatography, which are used for the separation of ionic 'bio-organic' types of compound and for high-molecular-weight materials[7].

7. Morris, C. J. O. R. and Morris, P. *Separation Methods in Biochemistry*, (1976) 2nd edn, Interscience, New York.

(a) General description

The separation is carried out using a column of the adsorbent packed into a glass tube (Fig. 4.16a) as a porous bed through which the mobile phase can flow. The mobile phase, generally known as the eluting solvent or eluent, is an organic solvent such as hexane.

The mixture to be separated is applied to the top of the column where it is adsorbed by the stationary phase. The eluent is then passed

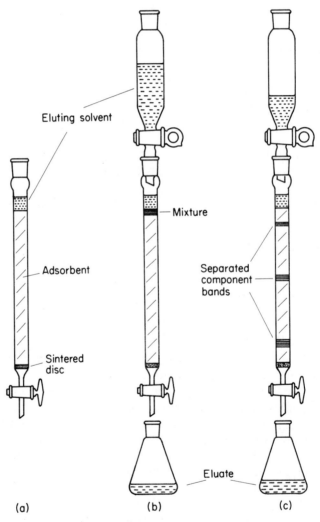

Fig. 4.16 Column chromatography (with gravity elution).

continuously through the column (Fig. 4.16b). Each component in the mixture is carried down the column by the mobile phase at a speed that depends on its affinity for the adsorbent. Ideally the mixture will separate into a number of discrete bands (Fig. 4.16c), which pass slowly down the column and eventually into a collecting vessel. Polar compounds such as alcohols (ROH), amines (RNH_2), or carboxylic acids (RCO_2H) are strongly adsorbed and move more slowly than less polar compounds such as aldehydes and ketones (RCOR'), ethers (R_2O) and hydrocarbons, which are less strongly adsorbed. The rate at which the bands move down the column can be controlled by adjusting the strength (polarity) of the eluting solvent (see Section 4.3.1).

It is usual to collect the eluate in batches (fractions) and to examine each by TLC (or GLC if appropriate) to determine which (if any) of the components is present. Appropriate fractions are then combined and the solvent is removed on a rotary evaporator to give the compound.

In all forms of column chromatography (see following sections) it is an essential part of good experimental practice to weigh the mixture before putting it onto the column and to weigh each component after separation. If this is not done you may 'lose' a compound in a complex mixture by leaving it undetected on the column.

(b) Choosing the method

There are many variations on the basic theme, which differ in the type of column used and in particular in the method used to drive the eluting solvent through the column. Historically, 'gravity elution' column chromatography (Fig. 4.16) was standard practice for many years, but it has been largely superseded over the last ten years or so by quicker and more effective methods such as flash chromatography, medium-pressure chromatography and dry-column flash chromatography.

In teaching exercises in the early stages it is usual to specify the method, adsorbent and eluting solvent to be used. However, in more advanced work it is essential to carry out a preliminary examination of the mixture by TLC. This will establish how many compounds are present, how easy or difficult they are to separate, and the best adsorbent and solvent (or solvent mixture) to use. The TLC plates will not necessarily have the same activity (see Sections 4.3.1 and 4.3.2) as the column but will still provide good guidance.

In general, compounds that can be resolved by TLC with an R_f difference of more than 0.10–0.15 can be separated by either 'flash' (p. 156) or 'dry-column flash' (p. 160). The 'medium-pressure' (p. 163) method is capable of higher resolving power and will produce good separations even when there is no clear separation of spots on TLC (HPLC must then be used for monitoring the separation). Traditional 'gravity elution' chromatography can be effective but tends to be tedious because of the low eluent flow rate, particularly when using long columns. Most separations can be carried out more effectively and more quickly with one of the previously mentioned methods.

SAFETY The dust from chromatographic adsorbents is harmful by inhalation. The operator must wear a dust mask, and all operations involving the loose adsorbent must be carried out in a fume-cupboard. The solvents used are usually volatile, flammable and to some degree toxic. Operators must take appropriate precautions to prevent ignition and inhalation (see Section 1.3). Operations in which solvents are left in open-topped containers should be carried out in a fume-cupboard.

(c) Flash chromatography

This technique was developed by Professor Clark Still[8] in the late 1970s as an effective and very rapid method for the routine separation of mixtures requiring only moderate resolution. It is capable of separating compounds that have an R_f difference (on TLC) of 0.15 or more.

The apparatus* used is shown in Fig. 4.17a. A number of columns of various diameters (but the same length) are required to accommodate different sample sizes (Table 4.3). The packing is a high grade of silica gel 60 (40–63 μm particle size), e.g. Merck type 9385. In operation the eluting solvent is driven down through the column by air or nitrogen pressure. The eluent flow rate is critical for good performance and the best results are obtained when the solvent level in the column above the packing falls at c. 2 inch (5 cm) min^{-1}. Using the conditions recommended below, the time required for elution is short (typically c. 15 min).

---

8. Still, W. C., Kahn, M. and Mitra, A. (1978) *J. Org. Chem.*, **43**, 2923. (We thank the American Chemical Society for permission to use material from this article, in particular Table 4.3 and Fig. 4.19. Copyright (1978) American Chemical Society.)

* Suitable apparatus is supplied by Aldrich Chemical Co. Ltd.

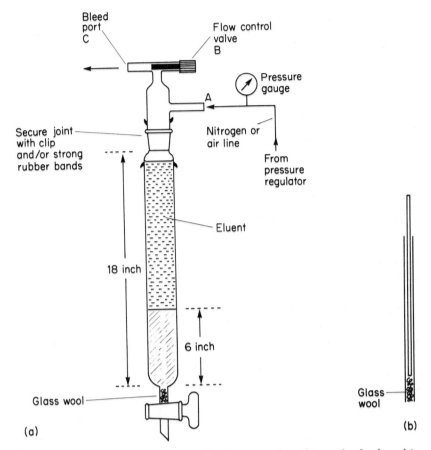

Fig. 4.17 (a) Apparatus for flash chromatography; (b) method of packing the glass wool plug.

(i) Selection of the eluting solvent

For general use the recommended solvent system is a mixture of petroleum ether (fraction b.p. 40–60°C) and ethyl acetate or, for more polar compounds, petroleum ether/acetone or dichloromethane/acetone. The procedure is as follows. Carry out a preliminary TLC examination to find the ratio of the two solvents that will produce an R_f of 0.35 for the desired compound. If several compounds run close on TLC, adjust the solvent to put the midpoint between the compounds at $R_f = 0.35$. If the compounds are widely separated adjust the R_f of the less mobile component to 0.35. Getting the solvent strength right is crucial to the success of the method and it is worth taking

Table 4.3 Various column sizes in flash chromatography

Column diameter (mm)	Volume of eluent [a] (ml)	Typical sample loading (mg)		Typical fraction size (ml)
		$\Delta R_f \geqslant 0.2$	$\Delta R_f \geqslant 0.1$	
10	100	100	40	5
20	200	400	160	10
30	400	900	360	20
40	600	1600	600	30
50	1000	2500	1000	50

[a] Typical volume required for packing and elution.

time over. If the percentage of the strong solvent required is very small (<2%), then it is best to reduce it to half for running the column.

(ii) Packing the column

Select a column of appropriate size (Table 4.3). Pack a small plug of glass wool into the tube at the base (Fig. 4.17a) using a piece of glass tube and a glass rod (Fig. 4.17b). Pour in a little coarse sand to make a smooth layer *c.* $\frac{1}{8}$ inch (2 mm) thick, and then − in one batch − pour in dry silica gel 60 (40−63 µm) to produce a column of length $5\frac{1}{2}$ − 6 inches (*c.* 15 cm). (See Safety note on p. 156.) With the stopcock open, tap the column gently on the bench (vertically) to consolidate the silica gel, and then add coarse sand to produce a layer $\frac{1}{8}$ inch thick on the top of the column. Pour in the eluting solvent carefully to fill the column completely. Fit the flow controller and secure all connections with joint clips and strong elastic bands (Fig. 4.17a).

SAFETY Take appropriate precautions for working with vessels under pressure.

Open the flow control valve B, attach the nitrogen line to A and set the pressure regulator on the cylinder to *c.* 7 psi. Nitrogen will flow vigorously out of the bleed port C. Close valve B until the pressure on the gauge reads about 7 psi. This will force the solvent rapidly through the silica, displacing the air in the column. It is essential to keep the solvent flowing without a break until all the air has been removed and the silica has cooled (otherwise the column will fragment − see p. 168). After this, adjust the flow controller so that the solvent level in the column is falling at the required 2 inches min^{-1} and note the pressure required. Continue until the solvent level is just above the top of the packing bed and then stop the flow by opening B fully and closing the stopcock on the column in that order. Do not allow the packing to run dry or the column will be ruined. The solvent used in this packing operation should be re-used to elute the column (sub-section (iv) below).

(iii)   Loading the sample

The preferred method is to apply the mixture as a fairly concentrated solution (*c.* 25% w/v) in the eluting solvent, as discussed in detail below. However, in many cases it will be found that the sample is not soluble enough in the initial eluent. In all forms of column chromatography it is important NOT to apply the sample either as a *dilute* solution in the eluent or as a solution in a stronger (more polar) solvent. In both cases this leads to band broadening and poor resolution. If the sample is not soluble enough in the eluent then it is best to pre-adsorb it on to the adsorbent as described below.

SAMPLE APPLICATION AS A SOLUTION. Prepare a solution of the sample in the eluting solvent (*c.* 25% w/v). Pipette it carefully onto the top of the adsorbent bed and then briefly pressurize the column to push the solution into the silica (take care NOT to push air into the packing bed). Wash down the walls with a little of the eluent and push the washings down into the silica.

PRE-ADSORPTION OF THE SAMPLE. In this technique the sample is pre-adsorbed onto *c.* 5 times its weight of the adsorbent and this is packed onto the top of the column. The procedure is as follows. Dissolve a known weight of the sample mixture in a good volatile solvent (e.g. dichloromethane) in a rotary evaporator flask. Add the adsorbent (*c.* 5 times w/w) and then remove the solvent on a rotary evaporator with gentle heating. The presence of the solid usually causes the mixture to 'bump' violently and it is therefore essential to incorporate an adapter (Fig. 4.18) containing a coarse glass sinter to prevent solid being lost into the evaporator duct. Eventually a free-running powder will be obtained. With the eluent level in the column a few centimetres above the adsorbent bed, add the pre-adsorbed sample via a funnel as a very thin stream, tapping the column gently so that it packs evenly onto the top of the packing bed. The addition must be done slowly, with the addition of more solvent if necessary, to avoid entrapment of air. Alternatively the pre-adsorbed sample can be suspended in a minimum volume of the solvent, and slurry-

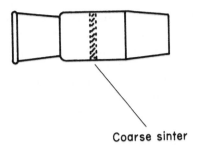

Coarse sinter

Fig. 4.18 Adapter with sinter for pre-adsorption.

packed (see p. 168). After packing, run the solvent level down and wash the walls as described in the subsection above.

(iv) Eluting the column

Fill up the column and reservoir with the eluting solvent. Attach and secure the flow controller and set the pressure at the value required to produce the desired flow rate (as determined in subsection (ii)). Collect fractions of the appropriate size (Table 4.3) using a rack of test tubes.

The separated components can be located by TLC as shown for example in Fig. 4.19, using a small glass TLC plate or more conveniently a strip of plastic-backed 'disposable' plate.

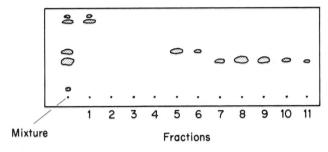

Mixture

Fractions

Fig. 4.19 TLC of fractions obtained in flash chromatography.

(d) Dry-column flash chromatography

This technique was developed by Dr L. M. Harwood at the University of Manchester in the early 1980s[9]. It requires no special equip-

9. Harwood, L. M. (1985) *Aldrichimica Acta*, **18**, 25. The method described here is based on this account and our own experience of the technique. (We thank the Aldrich Chemical Co. Inc. for permission to use material from this article.)

ment and provides separating power that is at least as good as flash chromatography and can be comparable with the resolution of an analytical TLC plate.

There are several major departures from conventional column methodology. The separation is carried out on TLC-grade silica (e.g. Merck Kieselgel 60), which has a much smaller particle size (*c.* 15 μm) than conventional column silica. The silica is packed into a standard P3 sintered glass filter funnel (Fig. 4.20) and the eluting solvent is drawn through by suction, using a water pump. The most unusual feature, however, is that the elution is carried out by adding the solvent in pre-measured portions, allowing the column to run 'dry' after each one before the next is added. It is very different but works remarkably well.

SAFETY The whole operation should be carried out in a fume-cupboard, away from sources of ignition.

(i) Apparatus

The apparatus is set up as for filtration (Fig. 4.20) using a parallel-sided filter funnel (P3 sinter) of size appropriate to the amount of sample (Table 4.4).

(ii) Packing the column

Fill the funnel up to the lip with dry TLC-grade silica (see Safety note on p. 156) and tap it gently to settle the powder and remove any voids.

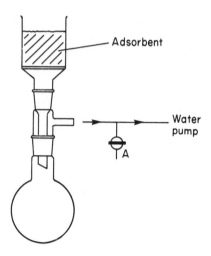

Fig. 4.20 Apparatus for dry-column flash chromatography.

Table 4.4 Various funnel sizes in dry-column flash chromatography

| Funnel size | | Silica weight (approx.) (g) | Sample weight[a] (mg) | Solvent fraction (ml) |
Diameter (mm)	Length (mm)			
30	45	15	200	10–15
40	50	30	500	15–30
70	55	100	1000–2000	20–50

[a] These weights are only approximate; lower loading may be appropriate for 'difficult' mixtures and higher loading for 'easy' ones.

Then apply suction, gently at first and then to full water-pump vacuum and press the silica down carefully with a flat stopper. Work around the circumference first and then in towards the centre to produce a totally level, well compacted bed. This produces a head space for the addition of the mixture and the eluent.

(iii) Choice of initial solvent

Carry out preliminary TLC experiments to find a solvent mixture that will give the fastest running component of interest an R_f of c. 0.2* and use this as the initial solvent. For compounds of moderate polarity it is usual to use hexane (or light petroleum) as the weak solvent with ether or ethyl acetate added to produce the required strength (see Section 4.3.1 (a)). No solvent is disfavoured and combinations of hexane (or pentane), ether, ethyl acetate or methanol are adequate for the full range of sample polarity.

(iv) Pre-elution

Under vacuum, pre-elute the column with the initial solvent, adding it gently to avoid breaking up the surface of the silica. If the packing has been done correctly the solvent front will be seen to descend in a horizontal line. If channelling occurs the column must be sucked dry and repacked. During pre-elution keep the surface covered with solvent until it is seen to pass into the receiver. Then allow the silica to 'dry' under suction – until the solvent flowing off the column has diminished to a few drips.

* This depends on the activity of the TLC plates; it may require a few preliminary experiments to establish the best correlation for your system.

(v) Loading the sample mixture

Ideally the mixture should be loaded onto the column as a concentrated (c. 25% w/v) solution in the initial solvent. (If it is not soluble
enough, follow the 'pre-adsorption' procedure in the next paragraph.)
With tap A (Fig. 4.20) open, pour the solution evenly onto the surface of the silica and then close tap A. When the column has been
sucked 'dry', commence elution (next subsection).

If the mixture is not soluble enough in the initial solvent then it
should be pre-adsorbed onto TLC silica as described on p. 159. Spread
the powder evenly over the compacted adsorbent (after pre-elution)
and press it down – under suction – carefully and evenly as in the
original preparation of the column.

(vi) Elution

Using a rack of test tubes make up the eluting solvent in the fraction size recommended in Table 4.4. A gradient of increasing solvent
strength produces the best results. Start with two or three tubes of the
initial solvent and then in each successive tube increase the proportion of the strong solvent by c. 5–10%, e.g. using a 10% increment,
if one tube contained 20% strong solvent the next would contain
22% and so on.

When these have been prepared pour the first one evenly onto the
top of the silica – with the vacuum line connected. Wait until the
column has been sucked 'dry' and then open tap A briefly, remove
the flask, and pour the eluate back into the test tube. Do the same for
each portion of solvent, allowing the column to be sucked 'dry' after
each addition. The components are located by TLC examination of
the fractions as described for flash chromatography. With practice,
separations of the same efficiency as TLC are easily possible. The
technique is not suitable for compounds that are easily oxidized.

(e) Medium-pressure liquid chromatography

The technique described here is based on that developed by Professor
A. I. Meyers[10]. Medium-pressure liquid chromatography (MPLC)
avoids many of the problems of gravity elution by using a pump to
deliver the solvent. This allows the use of long columns and smaller
column packing particles, which give higher separating power, with-

10. Meyers, A. I., Slade, J., Smith, R. K., Mihelich, E. D., Hershenson, F. M., and
Liang, C. D. (1979) J. Org. Chem., 44, 2247. (We thank the American Chemical
Society for permission to use material from this article. Copyright (1979) Ameri-
can Chemical Society.)

out the penalty of long analysis time. It is a much more powerful method than either flash or dry-column flash chromatography and is capable of separating mixtures for which there is no clear separation of spots on analytical TLC. It can be usefully combined with either of the 'flash' methods for the separation of complex mixtures, using the 'flash' technique for a quick preliminary separation and MPLC for the further separation of 'difficult' compounds. It is economical of adsorbent since the main column can be re-used many times after cleaning by back-flushing between separations.

(i) Apparatus

The apparatus is shown in Fig. 4.21. The columns are made of glass with solvent-resistant end fittings and are packed with silica gel 60 (40–63 μm), e.g. Merck type 9385, or alumina of similar quality. The packing procedure is described in subsection (vi) below. The size of the main separating column depends on the scale of operation and the ease or difficulty of the separation to be carried out. Columns 1000 × 15 mm are suitable for small-scale separations (up to c. 2–3 g) while 1000 × 25 mm columns can be used for samples up to c. 10 g.

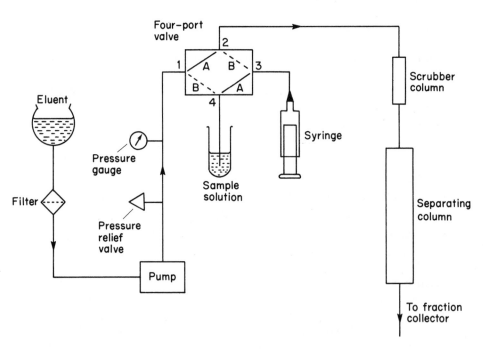

Fig. 4.21 Apparatus for medium-pressure chromatography.

These figures are only a rough guide since they may be exceeded for easy separations while difficult mixtures may only be resolved at very low column loadings. The small scrubber column (200 × 15 mm) is used to protect the main column from contamination by the highly polar 'tarry' materials found in many reaction mixtures and will therefore need to be repacked fairly often.

The pump (either a piston or a Teflon-coated diaphragm type) should be capable of delivering solvent at $5-50$ ml min^{-1} at pressures up to 100 psi. The pressure relief valve should operate at 100 psi.

The four-port valve has two positions: A, which connects the pump to the column and the syringe to the sample container: and B, which allows the sample to be injected onto the column. The injection syringe $(5-20$ ml) is preferably of the 'gas-tight' type.

The use of a mechanical fraction collector is virtually essential. Any type capable of collecting 25 or 50 ml fractions will suffice. (See safety notes on pp. 156, 158).

(ii) Selection of the eluting solvent

The solvent system is usually based on light petroleum (fraction b.p. $40-60$°C) as the weak solvent mixed with a strong solvent such as ether or ethyl acetate as required (see pp. 171, 172). Before the column separation is attempted the mixture should be examined by TLC to find the ratio of the two solvents that will give a R_f value of $0.2-0.3$ for the compounds to be separated (the lower value for difficult separations). Some experience is necessary to select the optimum solvent strength on this basis since it depends on the relative activity of the column and plate. For mixtures containing compounds with a wide polarity range the strength of the solvent will of course be increased as the separation proceeds, as is usual with gravity elution.

Having selected the initial solvent, use it to flush out the system – including the scrubber column only if the pre-adsorption method of sample application is NOT to be used (see next subsection).

(iii) Loading the sample

The preferred method is to inject the sample onto the column as a concentrated solution (c. 25% w/v) in the initial eluting solvent. As in all column chromatography the application of the sample dissolved in a very polar solvent should be avoided since this inhibits the initial sample adsorption onto the packing bed and leads to band broadening and poor resolution. If the sample is only adequately soluble in highly polar solvents then it is better to pre-adsorb it onto silica (p. 159) and dry-pack this into the top end of the scrubber column. See subsection (vi) below for the packing method.

When the sample is loaded as a solution it is injected via the two-way valve (Fig. 4.21) as follows. With the valve in the A position draw the solution into the syringe, then stop the pump, change the valve to the B position and inject the sample (injection is easier if the coupling between the scrubber column and the main column is disconnected during the operation).

If the pre-adsorption method is used the loaded scrubber column (dry-packed) is connected into the system after the main column has been flushed with the initial eluent. The air in the scrubber column is rapidly displaced from the system as the solvent is pumped through under pressure.

(iv) Elution

The flow rate used for elution should be the fastest that will allow resolution of the mixture and will depend on the difficulty of the separation. Minimum rates are c. 5 ml min^{-1} for the narrow (15 mm) columns and c. 20 ml min^{-1} for the wide (25 mm) columns. The location of the separated components in the fraction collector is done by TLC as described for flash chromatography (subsection (c) above).

(v) Back-flushing the columns

The columns can be used many times if they are cleaned up by removing all residual material by back-flushing between analyses. The scrubber column will need replacing more often depending on its colour.

After concluding the separation the apparatus should be connected as in Fig. 4.22 by coupling port 4 to the main column exit and port 3 to a waste bottle. With the valve in the B position flush the system with ethyl acetate and pump through about an additional 300 ml. The system should then be forward-flushed with the eluting solvent for the next separation or with a non-polar solvent if the columns are to be stored.

(vi) Column packing

Columns are dry-packed by the tap—fill method in which small portions of the adsorbent (enough for 3 mm column height) are added and consolidated by repeatedly tapping the column vertically on a hard surface. With some makes of column a recessed wooden block is necessary to avoid damage to the column end fittings during the tapping. With this tap—fill technique it is quite easy to achieve 900—1000 plates for the short (250 mm) column and 3000—4000 for the long (1000 mm) columns — these measurements were made using the

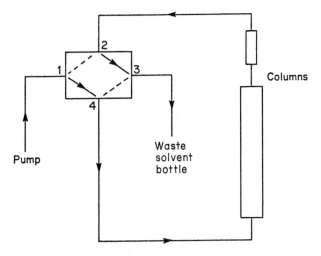

Fig. 4.22 Back-flushing the columns in MPLC.

columns in an analytical rather than a preparative mode with small sample sizes, on-column injections and low flow rates. In normal preparative use these efficiencies will not be realized because of column overloading and consequent band broadening but such measurements provide a useful check on the quality of the packing operation and can easily be made if an HPLC detector system is available (see Fig. 4.11). We have had much greater success with the tap–fill method than with the alternative tamping method of dry packing.

When the packing is complete the columns are connected for back-flushing as shown in Fig. 4.22 and solvent – usually light petroleum – is pumped through until all air is expelled. The column is then ready for use. When stored the columns are left wet with a non-polar solvent and capped to prevent ingress of air.

(f) Gravity elution chromatography*

In this technique the eluting solvent is supplied from a reservoir, usually a dropping funnel, attached to the top of the column (Fig. 4.16). The advantage is simplicity; there is little to go wrong in a mechanical sense. The major disadvantage is that the eluent flow rate becomes very slow when long columns or small particle size adsorbents are used. It is, however, useful for relatively easy separations, which can be carried out quickly on short wide columns.

* See Safety note on p. 156.

(i) The column

The column is a glass tube with a stopcock at one end and a socket at the other to take the solvent reservoir. The adsorbent is retained either by a sintered disc fused into the column (Fig. 4.16a) or, lacking that, by a plug of glass wool.

(ii) The adsorbent

Both alumina and silica gel can be used (particle size 50–200 μm). Alumina is generally preferred, not because silica gel is less effective at separating mixtures but because it has some specific problems that make it difficult to use in gravity columns. The major problem is that silica gel columns are susceptible to 'cracking' – the development of voids in the adsorbent bed. This tends to occur when the solvent strength is changed, particularly if the change is too abrupt, and results from the heat generated by solvent – adsorbent interaction. If silica is to be used, it is therefore important to use isocratic (constant-solvent-strength) elution or to change the solvent strength very gradually. Alumina is available in neutral, basic and acidic grades – all activated to Brockmann grade I (see Section 4.3.2). 'Neutral' is suitable for most types of compound, while 'basic' and 'acidic' adsorb organic acids and bases respectively very strongly. All types, however, when highly activated, can catalyse some remarkable 'on-column' reactions with sensitive compounds. It is important always to check that what comes off the column is what went on (TLC, NMR). The adsorbent should be deactivated to the required level (p. 175) before the column is packed. To achieve a comparable separation its activity should be higher than that of the TLC plate (usually II or III) used to establish the separation conditions.

The amount of adsorbent required depends to some extent on the difficulty of the separation; easy mixtures require an adsorbent: sample ratio (w/w) of about 50:1 for alumina and 75:1 for silica, while difficult mixtures may require a ratio of up to 300:1.

(iii) Packing the column

The column is packed using a mobile slurry of the adsorbent and the initial eluent (see subsection (v) below). The procedure is as follows. Partly fill the column with the initial solvent and then add the slurry in a steady stream via a funnel. Allow solvent to run out of the tap to accommodate the adsorbent and tap the column gently with a piece of heavy-wall rubber tubing to help the adsorbent to settle. Do not allow the solvent level to drop below the top of the adsorbent bed.

(iv) Loading the sample

AS A SOLUTION. The simplest method is to load the sample onto
the column as a concentrated solution in a little of the initial eluting
solvent (not a stronger solvent). The difficulty often encountered is
that the mixture is insoluble or only slightly soluble in the solvent. If
that is the case then the preadsorption method discussed on p. 159
should be used.

When it is soluble the procedure is as follows. Run off the eluent
from the column until it is just above the adsorbent bed. Add the
sample solution carefully, using a pipette, and then run off more solvent until the level is again just above the adsorbent. Wash down the
walls of the column carefully with a little of the initial solvent and
again run this down to just above the adsorbent. Fill up the column
carefully with the initial eluent, taking care not to disturb the top of
the packing, and commence elution.

It is common practice to put a thin layer of coarse sand on the top
of the adsorbent to prevent disturbance when filling the column with
the eluent.

(v) Eluting the column

Elution is carried out using organic solvents or solvent mixtures of
carefully controlled strength (see pp. 171, 172). It is usual to use light
petroleum (fraction b.p. 40–60°C) as the weak solvent, admixed with
ether or dichloromethane or some other strong solvent to produce
the strength required. The initial solvent for column packing is usually light petroleum or a solvent mixture shown by TLC (on plates
activated to the same level as the column packing) to produce an R_f
of 0.1–0.2 for the fastest moving component in the mixture. As the
chromatography progresses, the strength of the eluent can be progressively increased as appropriate by increasing the proportion of
the strong solvent. When eventually the proportion of the strong solvent (e.g. ether) reaches 100%, the strength can be further increased
by the progressive admixture of a yet stronger one (e.g. ethyl acetate)
and so on. Some guidance as to what is needed for each component
can be obtained by preliminary TLC experiments.

(vi) Fraction collection

Unless the compounds are coloured, it will not be obvious when each
component appears in the eluate. It is usual to collect equal-sized
'fractions' of eluate as the chromatography proceeds (see Table 4.3
for guidance on size of fraction) and to examine them by TLC (see
for example Fig. 4.19) or GLC to locate the compounds.

4.3 APPENDICES

4.3.1 General principles of chromatographic separation

All forms of chromatography are based on a system of two phases, one stationary and one mobile. This is best illustrated with examples.

(a) Liquid–solid chromatography

Sometimes this is known as adsorption chromatography because the stationary phase is a solid adsorbent (usually silica gel or alumina) in the form of small particles. The mobile phase (eluent) is an organic solvent such as hexane or ether. The adsorbent can be packed into a column (Fig. 4.16a) or spread as a thin layer on a backing plate (Fig. 4.1a). During operation the eluent flows through the adsorbent, either down the column or up the plate. A compound A loaded onto the adsorbent will distribute itself in a dynamic equilibrium between the adsorbent and the eluent. Its distribution ratio k'_A will depend on its relative affinity for the adsorbent and the eluent:

$$k'_A = n_S^A/n_M^A$$

where n_S^A is the total number of molecules of A adsorbed onto the solid and n_M^A is the total number dissolved in the eluent.

As the eluent flows through the adsorbent it will carry the sample along as a slowly moving band. The bigger the fraction of A in the eluent the faster it will pass through the adsorbent and, conversely, the more strongly A is adsorbed onto the solid the slower it will move. Thus the rate at which A moves depends on k'_A.

If a mixture of two different compounds (A and B) is loaded onto the top of the column there is a good chance that they will have different affinities for the mobile phase and the adsorbent, i.e. k'_A will be different from k'_B, and hence the two bands will move at different rates. Obviously the bigger the difference between k'_A and k'_B the greater the separation between them. The ratio k'_A/k'_B is known as the separation factor α.

The rate at which the bands pass down the column or up the plate can be controlled by varying the properties of the adsorbent (adsorbent activity – see below) and/or the properties of the eluting solvent (the solvent strength – see below).

ADSORBENT ACTIVITY. There are only two adsorbents in common use as stationary phases in liquid–solid chromatography: silica gel ($SiO_2 \cdot xH_2O$) and hydrated alumina (Al_2O_3). Both of these materials rely for their chromatographic properties on the presence of active adsorption sites on the surface, which can interact with organic molecules. Generally speaking the more polar the organic molecule the more strongly it will be adsorbed. (Functional groups with dipolar character, e.g $C = O$, $C \equiv N$, OH, NH_2, CO_2H etc., give an organic molecule polar character.) However, the active sites can also adsorb water molecules and these bind so strongly that the site is effectively blocked off and the surface is deactivated. The activity (adsorbing power) can therefore be controlled by adjusting the water content of the adsorbent. The activity is quantified by the Brockmann

Table 4.5 Solvent strengths (ε° values) of some chromatographic solvents[1,2]

Solvent	ε°
pentane (C_5H_{12})	0.00
hexane (C_6H_{12})	0.01
cyclohexane (C_6H_{12})	0.04
carbon tetrachloride (CCl_4)	0.18
toluene ($C_6H_5CH_3$)	0.29
2-chloropropane ($CH_3CHClCH_3$)	0.29
diethyl ether ($C_2H_5OC_2H_5$)	0.38
chloroform ($CHCl_3$)	0.40
dichloromethane (CH_2Cl_2)	0.42
tetrahydrofuran (C_4H_8O)	0.45
1,2-dichloroethane (CH_2ClCH_2Cl)	0.49
dioxane ($C_4H_8O_2$)	0.56
ethyl acetate ($CH_3CO_2C_2H_5$)	0.58
dimethylsulphoxide (CH_3SOCH_3)	0.62
acetonitrile (CH_3CN)	0.65
1- and 2-propanol (C_3H_7OH)	0.82
ethanol (CH_3CH_2OH)	0.88
methanol (CH_3OH)	0.95

[1] The values given are for chromatography on alumina but a similar order is observed for silica.

[2] More extensive data and discussion of the properties of chromatographic solvents can be found in Snyder, L. R. (1968) *Principles of Adsorption Chromatography*, Marcel Dekker, New York; and Snyder, L. R. and Kirkland J. J. (1979) *Modern Liquid Chromatography*, 2nd Edn, Wiley-Interscience, New York, and Knox, J. H. *High Performance Liquid Chromatography*, Edinburgh University Press, Edinburgh.

scale I–V, where grade I is the most active. The practical methods for controlling and measuring the activity are discussed in Section 4.3.2.

SOLVENT STRENGTH. The other major factor that controls the rate at which compounds pass down the column is the strength or polarity of the eluting solvent. The stronger (more polar) the solvent the greater its affinity for the compound on the column and the faster that compound passes through. The solvent strength is quantified as its $\varepsilon°$ value (Table 4.5), which ranges from 0 for non-polar hydrocarbons like pentane to 0.95 for the highly polar methanol. It is necessary to control the solvent strength very carefully during chromatography and this is usually done by using various mixtures of a strong solvent such as ether ($\varepsilon° = 0.38$) with a weak solvent such as hexane ($\varepsilon° = 0.01$). See for example Fig. 4.23, which shows how solvent strength varies with composition for binary mixtures of hexane with various 'strong' solvents.

(b) Gas–liquid chromatography

The principles of operation are similar to those for liquid–solid chromatography described above but in this case the stationary phase is a liquid and the mobile phase is a gas. The selectivity arises from partition rather than adsorption. The liquid stationary phase is a high-molecular-weight involatile fluid, usually a heavy oil, gum or polymer (for details see p. 130). Since it is liquid there is a basic problem in keeping it stationary. This is achieved in two ways. In one method the liquid is spread out as a thin film on the surface of an inert granulated solid known as the support material (Fig. 4.24a) (usually a form of Celite – a diatomaceous earth of high surface area). This coated support is known as the column packing and it is packed into a coiled glass or metal tube (usually 2–4 mm internal diameter and between 1 and 5 m long) known as the GLC column. A large surface area of the liquid phase is thus exposed to the gaseous mobile phase, usually nitrogen, which is passed continuously through the column during use (Fig. 4.2a). Packed columns of this type are reasonably cheap to make, have a long lifetime and are used extensively.

Alternatively the stationary liquid phase can be coated as a thin film onto the inside wall of a capillary tube of very narrow bore (0.2–0.5 mm) (Fig. 4.24b). The capillary tubes were originally made of metal but fused-silica capillary columns are now in general use. These columns have the advantage that the resistance to gas flow is very low and they can therefore be made very long (up to 100 m). This great length makes them very effective and they can achieve

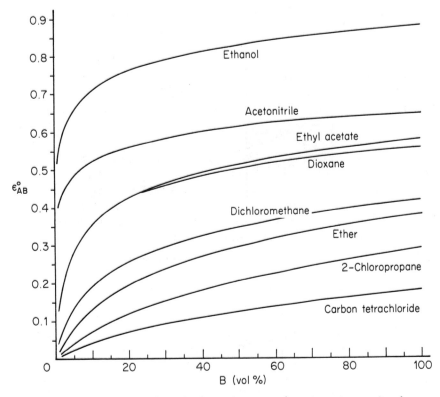

Fig. 4.23 Solvent strengths ($\varepsilon°$) for mixtures of various 'strong' solvents with hexane.

separations that are impossible on packed columns. They are, however, very expensive. Column efficiency is measured in theoretical plates (like distillation columns); a packed column would typically have *c*. 20 000 plates while a capillary column would have *c*. 250 000.

A compound when injected onto the column will partition itself in dynamic equilibrium between the stationary and mobile phases in proportions that depend on its volatility and its affinity for the stationary liquid phase. Each compound, for a particular operating temperature and liquid phase, will have its own distribution ratio (k', see p. 170). Since the mobile phase is flowing continuously through the column, the injected compound will pass as a band along the column at a rate that depends on k' (see p. 170). If a mixture of two compounds with different k' values is injected the more volatile one will partition with a higher concentration in the gas phase and so will pass along the column more rapidly than the less volatile one. Ideally complete separation of the two bands will occur and each compound

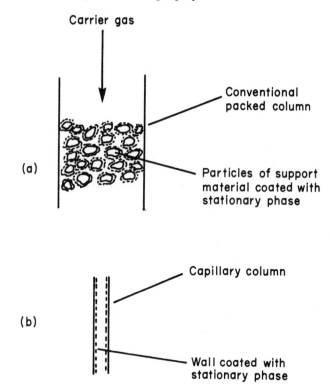

Fig. 4.24 Gas–liquid chromatography: (a) packed column; (b) capillary column.

will be eluted separately from the end of the column into a detector, which will produce a peak on a chart recorder trace, e.g. Fig. 4.2a.

(c) Liquid–liquid chromatography

The basic principle is the same as for GLC but in this case both the mobile and stationary phases are liquids. This would seem to pose considerable experimental problems in immobilizing the stationary liquid phase and preventing it being washed away by the mobile one. However, this difficulty has been overcome in the last few years by the development of bonded liquid phases. In these the stationary liquid phase is spread out as a thin film on the surface of an inert granulated solid – as in GLC – but the difference here is that the molecules of the liquid phase are chemically bonded to the surface of the particle and so cannot be washed off. The most common types use a spherical silica particle (Fig. 4.25) with a long-chain hydrocarbon

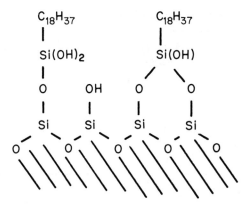

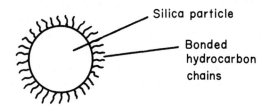

Fig. 4.25 Bonded stationary phase for LLC.

such as octadecane ($C_{18}H_{38}$) as the stationary liquid, bonded on at one end of the chain. The nature and polar properties of the bonded liquid can be changed by incorporating polar groups (e.g. cyano, amino, diol) into the hydrocarbon chain. These bonded materials can be packed into columns, where they are very important in HPLC, or coated onto plates for TLC. The operational methods are much the same as for liquid–solid chromatography but use different solvent systems as eluents.

4.3.2 Control of adsorbent activity

The activity of the adsorbent is controlled by adjusting its water content (p. 170). The activity is quantified as the Brockmann scale I–V[11]. The best way to produce a particular level of activity is first to drive off all the water by heating the adsorbent in a muffle furnace. Alumina requires 3 h at 500°C, and silica gel 16 h at 400°C to reach

11. Brockmann, H. and Schodder, H. (1941) *Chem. Ber.*, **74B**, 73.

Table 4.6 Water deactivation of silica and alumina

Water[a] (%) in silica	Activity grade	Water[a] (%) in alumina
0	I	0
5	II	3
15	III	6
25	IV	10
38	V	15

[a] By weight: X water + (100-X) adsorbent.

activity I. This material is then deactivated to the required level by adding water as indicated in Table 4.6.

The deactivation is carried out by placing the adsorbent, with the required amount of water, in a sealed container and mixing thoroughly for *c.* 5 h. On a large scale this can be done by rotating a glass or metal container on motor-driven rollers, and on a small scale in a closed rotary evaporator. When prepared, adsorbents should be kept in sealed containers to prevent adsorption of water from the atmosphere.

The Brockmann activity can be determined by finding the R_f values of various azo dyes (Table 4.7).

Table 4.7 Determination of Brockmann activity. R_f values on silica or alumina using 20% toluene/80% light petroleum as eluent.

| Dye | Activity | | | |
	II	III	IV	V
azobenzene	0.59	0.74	0.85	0.95
p-methoxyazobenzene	0.16	0.49	0.69	0.89
p-aminoazobenzene	0.00	0.03	0.08	0.19

4.3.3 Preparation of TLC plates

LARGE PLATES. These are coated using proprietary spreading equipment. For good results it is important that the plates are thoroughly degreased with detergent before coating. The spreader requires the use of a slurry of the adsorbent (10–30 μm particle size) in water (2 ml water per gram of adsorbent). As a guide 50 g of dry adsorbent is required to coat a plate area 100 × 20 cm to a thickness of 0.25 cm. After coating, the plates are left to dry on the spreader until the layer is set and are then transferred to a special rack and heated in an oven at 120–130°C for 4 h. This will produce plates of Brockmann activity II–III. The plates must be stored in a cabinet over silica gel.

MICROSCOPE SLIDES. These can be coated by dipping them in pairs, held back to back, into a 1:1 slurry of the adsorbent and methanol, quickly removing them, and wiping the long edges with the thumb and forefinger (wear a rubber glove). The coated plates are separated, placed horizontally on a suitable rack, and activated by heating in an oven at *c.* 120°C for 30 min. The thickness of the layer depends on the composition of the slurry. It can be conveniently kept in a small wide-necked screw-cap jar and used as required.

5 Preparation of samples for spectroscopy

The content of this chapter is strictly limited to that indicated in the title and is not concerned with spectroscopic theory or the interpretation of spectra. It is assumed that the reader understands the basic theory of each technique and the operating principles of the instruments. Operating instructions for infra-red (IR) and ultraviolet (UV) instruments are not given, as these vary and are best taught by demonstration.

5.1 INFRA-RED

Several methods of sample presentation are in regular use. The ones most popular in organic laboratories – 'liquid films' and 'mulls' – are discussed in subsection (a), and there are brief descriptions of the use of KBr discs and solutions in subsections (b) and (c) respectively.

(a) Liquid films and mulls

The simplest method involves the use of two rock salt plates (usually *c.* 25 mm diameter × 4 mm thick), which have been polished to produce optically flat surfaces. The material to be examined is simply sandwiched as a thin film between the two plates (Fig. 5.1). For liquid samples a thin film of the neat liquid is used, but solids must first be ground up with an inert liquid to produce a paste-like fine suspension known as a 'mull'. Sample preparation and loading the plates are discussed in more detail below. The plates are expensive and fragile and must be treated with great care. Obviously, being made of salt, they are water-soluble and the surfaces are rapidly damaged by putting wet samples or fingers on them. They should be handled – by the edges only – using rubber gloves, tongs or tweezers, and stored in a small desiccator over self-indicating silica gel. When loading the

Rock salt plates

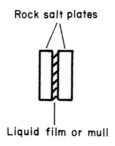

Liquid film or mull

Fig. 5.1 IR plates for liquid films or mulls.

sample onto the plates, put them on a piece of clean filter paper or a tissue (not directly onto the bench).

LIQUIDS. Place one drop of the neat liquid on one of the plates and then put the second one on top to trap it and spread the film.

SOLIDS. Place a little of the solid (*c.* 5 mg) in a small agate mortar and grind it to a fine powder. Then add a small drop of Nujol* (or other mulling liquid) and grind the mixture further to produce a smooth creamy paste. Judging the correct amount of Nujol takes practice; most people use far too much at first, you only need the smallest amount that will give a smooth paste. After grinding thoroughly, scrape the paste out of the mortar using a small spatula and spread it in a patch near the centre of one of the plates. Put the second plate on top and squeeze it down gently to spread the mull out into a thin film.

The quality of the mull has a marked effect on the quality of the spectrum. Too little Nujol gives a thick mull, which is difficult to spread and may produce broad, poorly resolved peaks in the spectrum. Too little grinding also produces poor resolution. Too much Nujol gives a spectrum dominated by the three Nujol peaks with only weak sample absorptions.

When the plates have been loaded, place them in whatever carrier is appropriate for your instrument and place the carrier in one of the IR beams (usually the nearer one). Most plate carriers produce some attenuation of the beam and it is necessary to balance this by putting a duplicate carrier in the other beam.

---

* Nujol is a viscous mixture of saturated hydrocarbons and gives IR absorptions at 2950, 1450 and 1375 cm^{-1}, which of course mask the sample absorptions at these frequencies. An alternative mulling agent that lacks C–H bonds is hexachlorobutadiene.

When the spectrum has been run the plates should be washed immediately. Separate them using tongs or tweezers and wash each with a jet of Arklone or dichloromethane from a wash bottle.

SAFETY Do this over a waste disposal bottle, preferably in a fume-cupboard

---

Dry the plates with a tissue and store over silica gel.

(b) KBr discs

The second most popular method for solid samples is to grind them with potassium bromide powder and then to compress the mixture at high pressure (10 000–15 000 psi) to produce a solid transparent disc. Typically this requires 1–2 mg of the sample plus c. 100 mg of potassium bromide* (the total amount of material required depends on the size of the die on your press but it should contain about 1% by weight of the sample).

The materials should be ground together as finely as possible in an agate mortar, or better, in a small ball mill. This should be done as quickly as possible to minimize the absorption of atmospheric water by the potassium bromide. The mixture is then transferred to the die of the hydraulic press and compressed. The disc produced should ideally appear transparent to visible light. However, translucent or even quite opaque discs can sometimes produce good spectra. Opacity of the disc is usually caused by adsorbed water or by using too much sample.

Obviously this process takes longer than making a mull and requires more equipment, but it offers the advantage that no absorptions from the mulling agent (or solvent) are present to mask sample absorptions.

(c) Solution spectra

Both solid and liquid samples can be examined as solutions. The major practical merit of using this technique is that it is non-destructive and valuable samples can be recovered. The solution is placed in a cell (Fig. 5.2) with rock salt windows (see subsection (a) for precautions in handling) whose path length can be adjusted by means of spacers over the range 0.015–1.0 mm. The thinnest spacers (0.015–

---

* Special IR-quality, dried at 110°C and stored in a desiccator.

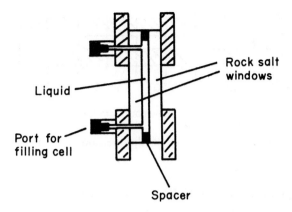

Fig. 5.2 IR liquid cell.

0.05 mm) give a path length suitable for neat liquids, while for solutions it is usual to use path lengths in the range 0.1–1.0 mm. Typically, a cell of path length 0.5 mm has a volume of 0.3 ml and, for a compound with strong absorptions, requires a solution concentration of *c*. 0.03 M. The cell is filled using a syringe, raising one end slightly during filling to ensure that all the air is expelled. After use, the cell should be washed out immediately using dry solvent (usually dichloromethane or Arklone), blown dry with clean air or nitrogen, and stored in a desiccator over silica gel.

SOLVENTS. The solvent used should have as few IR absorptions as possible, particularly in any region of interest for the sample. None are completely transparent over the whole range (660–4000 cm^{-1}) but carbon tetrachloride and carbon disulphide have few absorptions and are the solvents most often used.

SAFETY Both of these solvents are toxic by inhalation and carbon-disulphide has a very low flash point.

The clear windows for these solvents (shown in Fig. 5.3) overlap, so the use of both will give full coverage. The solvent absorption peaks are compensated by using a matched cell containing the pure solvent in the reference beam of the spectrometer. Note, however, that although this prevents the solvent peaks appearing on the IR spectrum of the sample, any sample peaks that would have appeared at these points will be masked so that they either do not appear at all or appear at a much reduced intensity.

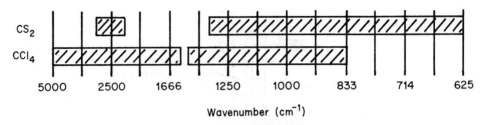

Wavenumber (cm⁻¹)

Fig. 5.3 IR 'windows' for carbon disulphide and carbon tetrachloride.

5.2 NUCLEAR MAGNETIC RESONANCE

An NMR spectrum (^{1}H or ^{13}C or other magnetic nucleus) is obtained using a solution of the sample in a non-interfering solvent (as far as that is possible). Most spectrometers accept sample tubes that are 5 mm diameter (Fig. 5.4a) and require c. 0.4 ml of solution, which fills the tube to a depth of c. 35 mm. Care is needed in the preparation of the solution since the quality of the spectrum obtained depends not only on the purity of the sample but also on the cleanliness of the tube, the purity of the solvent and in particular on proper filtration to remove suspended dust.

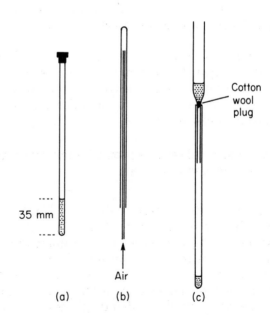

Fig. 5.4 Sample preparation for NMR spectroscopy.

(a) The NMR sample tube

These are precision items, made from thin-walled tubing, and are easily chipped or broken at the open end. Treat them with care, as they are very expensive. A tube should be thoroughly washed out with acetone or dichloromethane immediately after use, dried by blowing in clean air or nitrogen using a thin piece of metal tubing (Fig. 5.4b), and then capped to keep out dust in storage. Drying the tubes by baking in an oven is not recommended as they pick up dust if left open for long and it is said that baking can produce distortions. Immediate washing out is important as old solutions can decompose and polymerize, leaving residues that are hard to remove. In such cases a piece of pipe cleaner makes a useful 'brush', used with an appropriate organic solvent. If this is not effective, then consult your instructor; do not use abrasives or chromic acid.

(b) Preparing the solution

In most cases in teaching laboratory work you will be told how much sample and which solvent to use – if not then see subsections (c) and (d) below. The solution is conveniently prepared in a small (*c.* 3 ml) sample vial, which should be washed and oven-dried before use. Weigh in the sample, add 0.4 ml of the solvent using a 1 ml graduated pipette and tap the tube gently to mix the contents* and dissolve the sample. It is essential to filter the solution to remove dust or other suspended material. This can be done quickly and easily using the Pasteur pipette/cottonwool filter shown in Fig. 5.4c. To prepare the filter, tear off a tiny piece of clean dry cottonwool, crush it and pack it fairly tightly into the throat of the pipette using another pipette to tamp it down. For best results run a little clean solvent through the filter to wash out any loose fibres. Then place it in the neck of the NMR tube (Fig. 5.4c) and pipette in the solution using another Pasteur pipette (NOT the graduated pipette, which should be reserved for clean solvent only, to avoid risk of contaminating the stock solvent bottle). When filtration is complete, cap the tube and keep it vertical and away from light and heat until the spectrum is run.

---

* Some spectrometers require the presence of an internal standard, usually tetra-methylsilane $((CH_3)_4Si)$. If it is required for your particular instrument it will probably have been added to the stock bottle of NMR solvent (*c.* 0.5%). Check with your instructor and if you need to add a standard, do it at this stage.

(c) NMR solvents

Various perdeuteriated solvents are used, e.g. deuteriochloroform $(CDCl_3)$ or deuteriobenzene (C_6D_6). The use of deuteriated solvents is obviously essential for ^1H NMR spectroscopy so that solvent ^1H absorptions do not obscure the spectrum. However, less obviously, they are also required when running spectra of other nuclei (e.g. ^{13}C) as most instruments use the deuterium nucleus to provide a frequency 'lock' signal. A deuterium content of 99.5 at % or better is required for ^1H spectroscopy. The most widely used solvent by far is deuterio-chloroform $(CDCl_3$: $\delta(^1H)^*$, 7.25; $\delta(^{13}C)$, 76.9): it is a good solvent, unreactive to most functional groups (some amines react) and reasonably cheap.

SAFETY Chloroform is toxic and should be used in a fume-cupboard.

Other useful solvents (but much more expensive than $CDCl_3$) are shown in Table 5.1.

Solubility testing should of course be done using non-deuteriated solvents.

(d) Amount of sample

It is very difficult to give firm guidance here as this depends on the nucleus to be observed and the nature of the spectrometer. However, some rough guidelines for ^1H and ^{13}C are offered.

^1H NMR. Many routine spectra for teaching laboratory and project work are run on 60 MHz continuous-wave (CW) instruments and for these a spectrum with a good signal-to-noise ratio can be obtained using about 0.1 mmol of sample (say 30–50 mg for molecular weights (MW) up to 500). Using a more sophisticated Fourier Transform (FT) instrument, spectra can be obtained using much smaller samples (1 mg or less), but *c.* 20 mg should be used if available to keep the instrument time to a minimum.

^{13}C NMR. Because of low natural abundance and lower magnetic moment, ^{13}C NMR is *c.* 6000 times less sensitive than ^1H NMR.

---

* The ^1H shift is useful to know since this absorption may be seen in spectra run using very dilute solutions when the concentration of the residual *c.* 0.5% of protium-containing solvent is comparable with the solute concentration.

Table 5.1 Useful NMR solvents

Solvent	$\delta\,(^1H)$ [a]	$\delta\,(^{13}C)$
deuterio-		
-chloroform ($CDCl_3$)	7.25	76.9
-benzene (C_6D_6)	7.16	128.0
-toluene ($CD_3C_6D_5$)	1.99 and	19.2 and
	6.98–7.09	128.9, 137.5
-acetone (CD_3COCD_3)	2.05	29.8, 206.0
-dimethylsulphoxide (CD_3SOCD_3)	2.50	39.5
-trifluoroacetic acid (CF_3CO_2D)	11.30	115.7, 163.8
-water (D_2O)	4.8 (at pH 7, pH-dependent)	

[a] See footnote to p. 184.

Solutions should therefore be as concentrated as possible. Using a typical high-field FT instrument (90 MHz for ^{13}C) a sample size of 0.1 mmol requires about 1 h acquisition time to produce a fully proton-decoupled spectrum and a further 1 h to give two DEPT (distortionless enhancement by polarization transfer) spectra to determine the proton multiplicity of the carbons (i.e. CH_3, CH_2, CH or quaternary). The important relationship between acquisition and concentration is the following: if the concentration is halved, the acquisition time required goes up by a factor of 4. Thus a sample of size *c.* 0.025 mmol would require an overnight run on the spectrometer.

5.3 MASS SPECTROMETRY

This is a highly sensitive technique and spectra can be produced using only micrograms of material. However, sample handling is easier with larger amounts and you should submit a few milligrams if available. No sample preparation is needed as this is carried out where necessary by the instrument operator. The operator will need to know the approximate molecular weight of the compound and its m.p. or b.p. and it is helpful to give some indication of its thermal stability if known.

It is important that the sample is pure. Mass spectra obtained from impure compounds can be very misleading since the peaks produced by slightly more volatile impurities, even if present in only tiny amounts, may be recorded at higher intensity than those from the major component.

5.4 ULTRAVIOLET

The determination of a UV spectrum requires the preparation of a very dilute solution of the sample (of known concentration) in a solvent that is transparent to UV radiation. The method of making up the solution is described in subsection (d) below.

(a) Sample cells

A typical cell is shown in Fig. 5.5. Cells of 10 mm path length (volume *c.* 3 ml) are normally used but 1 mm and 40 mm cells are also usually available. These are precision items, made of silica, and are very expensive. Similar cells made of glass are used for visible light (vis.) spectra. It is important to avoid any abrasion or damage to the transparent walls of the cell. After use the cell should be washed out thoroughly with solvent and blown dry with clean air or nitrogen. If more powerful cleaning is ever required, use one of the proprietary glassware detergents followed by rinsing with distilled water and then ethanol.

NEVER use abrasive powders or test-tube brushes.

   When running the spectrum a cell containing the sample solution is placed in one beam of the spectrometer and a matched cell containing the solvent only is placed in the other beam.

(b) Solvents

These are usually saturated hydrocarbons (e.g. hexane) or simple alcohols (e.g. ethanol). They should be of special 'spectroscopic'

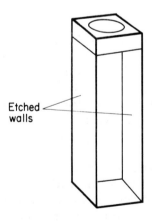

Etched walls

Fig. 5.5 Sample cell for UV/vis. spectroscopy.

quality, free from UV-absorbing impurities, which may be present in laboratory-grade solvents. Since the solutions required are usually dilute ($c.$ $5-50$ mg l^{-1}), solubility is rarely a problem.

(c) Concentration

The concentration required depends on the extinction coefficient of the sample (ε in the equation below). The spectrum you will obtain is a plot of absorbance (A) (or optical density) against wavelength. The absorbance is dependent on the concentration of the solution (c) and the path length of the sample cell (l), and ε is the proportionality constant that relates A to c and l at a particular wavelength, i.e.

$$A = \varepsilon c l$$

(It is usual to express c in moles per litre and l in centimetres; ε is then known as the molar extinction coefficient and is usually quoted without units although it is not dimensionless.) Thus for benzoic acid, $C_6H_5CO_2H$ (λ_{max}^{EtOH} 228 nm; ε_{max} 10 660), to obtain an absorbance value of 1.9 (utilizing almost the full height of the chart), using a cell of path length 1 cm, a solution of concentration 1.8×10^{-4} mol l^{-1} (0.022 g l^{-1}) would be required.

(d) Making up the solution

The direct preparation of the very dilute solutions required would require high accuracy in weighing out a tiny amount of sample, or the use of a large quantity of expensive 'spectroscopic'-grade solvent. To avoid this it is usual to prepare a stock solution of $c.$ $10-100$ times the required concentration and then to dilute it using pipettes and graduated flasks until the required concentration is reached. If the compound has a known UV spectrum then look it up (see references on p. 190) and use the equation in the previous subsection to calculate the concentration required. For example, to obtain the UV spectrum of benzoic acid (see subsection (c)), a solution of suitable concentration could be produced by making up a stock solution of benzoic acid (55 mg) in ethanol in a 25 ml graduated flask and – after homogenizing it – diluting 1 ml of this solution to 100 ml in a second graduated flask. If the compound is new then you will have to experiment with various successive dilutions of the stock solution – running a spectrum for each – until the required concentration is obtained.

The preparation of the stock solution and the dilutions should follow the normal procedures used in quantitative work and all pipettes and flasks must be absolutely free from organic contaminants and solvents, which could give spurious absorptions.

6 Finding chemical information

The use of the literature of organic chemistry is an essential complement to experimental work at all levels[1]. By 'the literature' we mean the enormous volume of information stored in our libraries on chemical reactions and the many millions of compounds that have been synthesized since organic chemistry first got under way in the middle of the last century.

The basic source of all information is the **primary literature**. New chemical knowledge is first reported by its discoverers in an 'original paper' in one of the hundreds of chemical journals that are published throughout the world in many languages. Most chemical libraries have complete runs of the more important of these journals from their beginnings up to the present day. Less well known journals can often be obtained on loan from central libraries as required. Access to this vast store of knowledge is best achieved by consulting *Chemical Abstracts* into which the contents of every paper, patent, etc. are abstracted and all the information grouped together under broad subject headings and indexed (see Section 6.4 below). Much of this information is subsequently collected together, correlated and discussed in an extensive range of reviews and monographs on selected topics such as particular classes of reactions or particular classes of compounds. Publications of this kind constitute the **secondary literature** and are invaluable for acquiring information about specific areas of chemistry when writing essays or preparing for a research project. In addition much of the physical and spectroscopic data, particularly on relatively simple compounds, has been collected together in a wide range of reference and data books.

In the future most searching of the literature will be done by computer scanning of data bases and abstracts. Much progress has been made already towards this end, but, for reasons of cost rather than practicability, we still rely heavily on manual methods.

[1] The reader is referred to Maizell, R.E. (1987) *How to Find Chemical Information*, 2nd Edn, Wiley Interscience, New York, for a more detailed coverage.

Let us look first at the different types of information required during a practical course. The first type relates to the physical properties of specific compounds, usually their melting or boiling points, and to details of their IR, UV, NMR or mass spectra. This information is usually required during exercises on the identification of unknown compounds. Often, final confirmation of structure depends on the comparison of the physical and/or spectroscopic properties of the unknown compound with those of candidate structures recorded in the literature. In the past, identification procedures have depended largely on the preparation of one or more crystalline derivatives of the unknown compound and comparison of their melting points with those published in the literature. This approach has become less widely used as spectroscopic methods of identification have become more effective and more extensively applied in undergraduate teaching, but it is still pertinent in some cases. The physical and spectroscopic data required for the identification of unknown compounds can often be obtained from reference books of various kinds (see Sections 6.1 and 6.2 below), but for less common compounds it is usually necessary to consult *Beilstein's Handbuch* (Section 6.3) or refer back to the original literature via *Chemical Abstracts* (Section 6.4).

The second type of information sought in practical work is concerned with the methods of preparation of particular compounds and with more general aspects of their reactivity. For example, in project work you may need to search the literature to find out the best way of preparing a particular compound for use as a starting material. You may also want to know whether the compound you need to make, or perhaps have just made, is 'new' in the sense that it has never been prepared before. This sort of information may be obtainable from *Beilstein's Handbuch* which provides a wealth of detail on the methods of preparation and properties of an enormous range of compounds with references to the original literature (see Table 6.1 for period of literature coverage). Its use is discussed in Section 6.3. However in most cases it will be necessary to utilise *Chemical Abstracts* and undertake a thorough search of the primary literature.

6.1 PHYSICAL PROPERTIES

Data on the melting and boiling points of relatively common organic compounds and on the melting points of their crystalline derivatives can be obtained from many books dealing with the identification of organic compounds such as Vogel's *Practical Organic Chemistry* in which compounds are classified by their functional group and listed in tables in order of increasing boiling or melting point. The most comprehensive set of data of this kind is found in the *CRC Handbook of Tables for Organic Compound Identification*. Information on a very much wider range of compounds can be obtained from *Dictionary of Organic Compounds*. This is a multi-volume handbook which lists compounds in alphabetical order and, most usefully, also gives references to methods of preparation together with details of derivatives where appropriate.

Useful physical data (m.p., b.p., density, refractive index, solubility) of a wide range of common organic compounds is available from the *CRC Handbook of Chemistry and Physics*.

The largest collection of data on organic compounds is presented in *Beilstein's Handbuch* (see Section 6.3). It gives information on some physical properties (m.p. or b.p. and solubility) as well as on methods of preparation and reactions.

6.2 SPECTROSCOPIC DATA

A general guide to structure determination by the joint application of UV, IR, NMR and mass spectroscopy (such as the excellent text *Spectroscopic Methods in Organic Chemistry* by Williams and Fleming) is an essential companion to all laboratory courses. Books of this type outline the principles of each spectroscopic method and provide examples of structure elucidation as well as tables of correlation data for interpreting the spectra of unknown compounds.

It is often necessary to find the spectra of particular compounds for comparison purposes. The most readily available reference collections are listed below. However it must be emphasized that these are far from comprehensive and it may be necessary to go back to the original literature to obtain the information required (see Sections 6.3 and 6.4).

UV   (i)   *The Sadtler Handbook of Ultraviolet Spectra* which contains the spectra of 46 000 compounds.

   (ii)   *Organic Electronic Spectral Data*, in which compounds are indexed by their empirical formula, gives absorption maxima together with literature references.

   (iii)   *UV Atlas of Organic Compounds* contains spectra of nearly 1000 compounds, cross-indexed by chromophoric group.

   (iv)   *Ultraviolet Spectra of Aromatic Compounds*, gives the spectra of 597 compounds.

IR   (i)   *The Aldrich Library of Infrared Spectra* contains over 12 000 spectra arranged in order of increasing molecular complexity. Fifty one chemical classes are represented, indexed alphabetically and by molecular formula. This is an excellent reference book for undergraduate use.

(ii) *Documentation of Molecular Spectroscopy* provides the most extensive set of data, elaborately cross-indexed on coded cards.

(iii) *The Sadtler Handbook of Infrared Spectra* is a collection of 55 000 spectra (prism and grating) with additions each year.

NMR (i) *The Aldrich Library of NMR Spectra* contains around 8500 proton spectra recorded on a 60 MHz instrument. It is excellent for undergraduate use.

(ii) *Atlas of Carbon-13 NMR Data* lists the chemical shifts and assignments of over 3000 compounds. Also available as a data base for computer retrieval.

(iii) *Carbon-13 NMR Spectra* contains c. 500 spectra.

Mass spectra

N.B. Caution must be exercised in identifying compounds using only the comparison of mass spectra with library data because there can be considerable instrument-dependent variation in the relative abundances of the fragment ions.

(i) The *Eight Peak Index of Mass Spectra* catalogues the eight most abundant ions for over 30 000 compounds.

(ii) *American National Bureau of Standards Data Base* is a library of over 42 000 spectra and is often used as a data base in the computers of modern mass spectrometers for comparison with acquired spectra.

(iii) *Lederberg's Computation of Molecular Formulas for Mass Spectrometry* is a reference work on high resolution mass spectral data and gives all the possible molecular formulae that satisfy a particular molecular weight when it is known to five or six decimal places.

6.3 BEILSTEIN'S HANDBUCH

This is a unique reference work (in German) covering all organic compounds reported in the literature up to 1959 and at present being expanded to provide coverage up to 1979, see Table 6.1. It provides information on methods of preparation, reactions and physical pro-

Table 6.1 The Series of *Beilstein's Handbuch* (4th edn)

Series	Abbreviation	Period of literature completely covered	Colour of label on spine
Basic Series	H	up to 1909	green
Supplementary Series I	E I	1910–1919	red
Supplementary Series II	E II	1920–1929	white
Supplementary Series III	E III	1930–1949	blue
Supplementary Series III/IV	E III/IV*	1930–1959	blue/black black
Supplementary Series IV	E IV	1950–1959	(to be
Supplementary Series V	E V	1960–1979	published from 1984/ 85 on)

* Volumes 17–27 of Supplementary Series III and IV, covering the heterocyclic compounds are combined in a joint issue.

perties. It is not difficult to use once the way in which the material is organized is understood.

The entire series consists of the main work (*Hauptwerk*, H) together with a number of supplements (*Ergänsungbande*, EI, EII, etc.) some of which are not yet complete, covering the periods listed in Table 6.1. The *Hauptwerk* is divided into acyclic (vol. no. 1–4), carbocyclic (vol. no. 5–16) and heterocyclic (vol. no. 17–27) compounds. Each supplement also comprises 27 volumes (or groups of volumes) and parallels that of the *Hauptwerk* in the arrangement of material. There is a General Subject Index (name index of the compounds dealt with) and a General Formula Index (index of empirical formulae) for the *Hauptwerk* and the first two supplements (EI, EII) which, together, cover the literature up to 1929. Thereafter, separate ones appear for each volume of the other supplements (EIII onwards). A particularly helpful feature of *Beilstein* is the System Number given to each compound (and those with closely related structures) which allows it to be found quickly in subsequent supplements under the same volume number (provided relevant new findings have been published in the literature). In addition, the page number on which the compound first appeared is repeated in bold print on the tops of the pages of the various supplements, enabling any additional information to be located easily.

When trying to locate a compound the simplest method is via the indexes. For the beginner the nuances of nomenclature often pose a problem and it is simpler and quicker to search in the General Formula Index (2nd edn) under the compound's molecular formula. In the formula index C and H come first, followed by the other elements present in alphabetical order. Let us suppose that we wish to identify a compound, $C_{10}H_{10}O$ (m.p. 41–42°C) that is known, from its spectroscopic and chemical properties, to be an α, β-unsaturated ketone. Two of the names listed under $C_{10}H_{10}O$ can be recognized as belonging to compounds of this class and these entries read as follows:

Benzylidenaceton 7, 364, I 192, II, 287

ω-Äthyliden–acetophenon 7, 368, I 194, II 290.

This means that both compounds occur in volume 7; the three page numbers that follow refer to the *Hauptwerk*, the first supplement (I) and the second supplement (II). Note that the volume number always remains the same.

In the individual entries, methods of preparation are given first, then physical properties (m.p., b.p., an indication of solubility, etc) then chemical properties, beginning with the action of heat, oxidizing and reducing agents. If a compound forms salts these are given at the end of the entry. It is useful to know that

F	= melting point
Kp	= boiling point (a subscript indicates pressure in mmHg)
leicht löslich	= easily soluble
schwer or wening löslich	= sparingly soluble
unlöslich	= insoluble

If all else fails there is usually a German–English dictionary in the library.

The entries for the two ketones mentioned above indicate that benzylideneacetone is indeed a solid melting at 41–42°C but ethylideneacetophenone is a liquid at ordinary temperatures and therefore excluded as a possibility.

The third supplement (E III) which covers the literature up to 1949 is complete, but does not as yet have cumulative indexes. For locating individual compounds in the third supplement, *either* look in the name or formula index provided at the end of each individual volume (volume 7 for the two examples above), *or* note the number of the page on which the compound appears in the *Hauptwerk*. This number is reproduced in the middle of the top of the equivalent page in each supplement. Thus benzylideneacetone can be found on page 1399 in volume 7 of the third supplement.

6.4 CHEMICAL ABSTRACTS

This work, published by the American Chemical Society, began in 1907 and covers the entire literature in the whole field of chemistry. It appears weekly and consists of short summaries in English of original articles or patents published anywhere in the world. Over 10 000 journals in many different languages are abstracted in a standard format, giving

(i)   the abstract number
(ii)   the title of the paper
(iii) the names of the authors
(iv)   the address of the authors
(v)   the abbreviated name of the journal in which the paper appears
(vi)   the year, volume of the journal, and page number of the journal, and
(vii) the language of the paper

There are annual indexes of subjects, authors, formulae and patents for each volume until 1962 when two volumes per year were issued, each with a separate index. From 1972, the *Subject Indexes* have been sub-divided into a *Chemical Substance* and a *General Subject Index*. At intervals of ten (more recently five) years the indexes to each volume are collected together and published as *Collective Indexes*. Those published so far are shown in Table 6.2 and cover the

Tabe 6.2 Indexes for Chemical Abstracts

Collective indexes	Volume numbers	Subject	Chemical substance	Author	Formula	Patent number
1	1–10	1907–1916		1907–1916		
2	11–20	1917–1926		1917–1926		1907–1936
3	21–30	1927–1936		1927–1936	1920–1946	
4	31–40	1937–1946		1937–1946		1937–1946
5	41–50	1947–1956		1947–1956	1947–1956	1947–1956
6	51–55	1957–1961		1957–1961	1957–1961	1957–1961
7	56–65	1962–1966		1962–1966	1962–1966	1962–1966
8	66–75	1967–1971		1967–1971	1967–1971	1967–1971
9	76–85	1972–1976	1972–1976	1972–1976	1972–1976	1972–1976
10	86–95	1977–1981	1977–1981	1977–1981	1977–1981	1977–1981
11	96–105	1982–1986	1982–1986	1982–1986	1982–1986	1982–1986

literature up to 1986. Thereafter, it is necessary to consult the bi-annual indexes.

Any search for a specific compound should begin with the latest indexes and progress back in time, at least until 1949, when *Beil-stein's Handbuch* provides direct access to selected data. The inexperienced searcher will find it easiest to start by using the Formula Index since this avoids the pitfalls of incorrectly naming the compound. Such problems however can be much reduced by using the Index Guide. This is produced with each Collective Index (from the 8th) and provides indexing notes, alternative names for compounds, cross-references and illustrative structural diagrams. Another invaluable aid designed to help the user locate a compound quickly is the *Ring Index*, which lists the systematic names of ring systems according to the number of rings in the compound as well as their size and the constituent atoms. For example, the following compound will be

classified as 5, 6, 6, 6 and C_4N-C_5N-C_6-C_6 and will appear in the *Ring Index* under 4-ring systems as a 11*H*-Indolo[3,2-*c*]quinoline.

BIBLIOGRAPHY

'Beilstein's Handbuch der Organischen Chemie', 4th edn., Springer-Verlag, Berlin, 1918–present. Multi-volume.

E. Breitmaier, G. Haas, and W. Voelter, 'Atlas of Carbon-13 NMR Data', Vols. 1 and 2, Heyden, London, 1979.

'Chemical Abstracts', American Chemical Society, Chemical Abstracts Service, Columbus, Ohio, 1907–present. 2 vols. per year with volume indexes.

'CRC Handbook of Tables for Organic Compound Identification', 3rd edn. compiled by Z. Rappoport. CRC Press Inc., Cleveland, Ohio, 1967.

'CRC Handbook of Chemistry and Physics', ed. R. C. West, CRC Press Inc., Cleveland, Ohio, Annual.

'Documentation of Molecular Spectroscopy', Butterworths, London, A collection of spectra on coded cards, elaborately cross indexed.

'Eight Peak Index of Mass Spectra', 2nd edn., UKCIS, The University, Nottingham, U.K. 1974.

'Dictionary of Organic Compounds', 5th edn., Chapman and Hall,

London, 1982. 5 vols. plus annual supplements and 2 vols. of indexes.

L. F. Johnson and W. C. Jankowski, 'Carbon-13 NMR Spectra, a Collection of Assigned, Coded and Indexed Spectra', Wiley, New York, 1972.

J. Lederberg, 'Computation of Molecular Formulas for Mass Spectrometry' Holden-Day, San Francisco, 1964.

C. J. Pouchert, 'The Aldrich Library of Infrared Spectra', 3rd edn., Aldrich Chemical Co., Inc., Milwaukee, WI, 1981.

C. J. Pouchert, 'The Aldrich Library of NMR Spectra', 2nd edn., Aldrich Chemical Co., Inc., Milwaukee, WI, 1983.

'Organic Electronic Spectral Data', Wiley, New York, 1960–present. Annual since 1972 (covers literature back to 1946).

'The Sadtler Handbook of Infrared Spectra', ed. W. W. Simons, Heyden, London, 1978.

'The Sadtler Handbook of Ultraviolet Spectra', ed. W. W. Simons, Heydon, London, 1978.

'UV Atlas of Organic Compounds', Butterworths, London, 1965.

A. Vogel, 'Textbook of Practical Organic Chemistry', 4th edn., Longman, London, 1978.

D. H. Williams and I. Fleming, 'Spectroscopic Methods in Organic Chemistry' 3rd edn., McGraw-Hill (U.K.), 1980.

Index